Undergraduate Lecture Notes in Physics

For further volumes:
http://www.springer.com/series/8917

Undergraduate Lecture Notes in Physics (ULNP) publishes authoritative texts covering topics throughout pure and applied physics. Each title in the series is suitable as a basis for undergraduate instruction, typically containing practice problems, worked examples, chapter summaries, and suggestions for further reading.

ULNP titles must provide at least one of the following:

- An exceptionally clear and concise treatment of a standard undergraduate subject.
- A solid undergraduate-level introduction to a graduate, advanced, or non-standard subject.
- A novel perspective or an unusual approach to teaching a subject.

ULNP especially encourages new, original, and idiosyncratic approaches to physics teaching at the undergraduate level.

The purpose of ULNP is to provide intriguing, absorbing books that will continue to be the reader's preferred reference throughout their academic career.

Series Editors

Neil Ashby
Professor Emeritus, University of Colorado Boulder, CO, USA

William Brantley
Professor, Furman University, Greenville, SC, USA

Michael Fowler
Professor, University of Virginia, Charlottesville, VA, USA

Michael Inglis
Professor, SUNY Suffolk County Community College, Selden, NY, USA

Elena Sassi
Professor, University of Naples Federico II, Naples, Italy

Helmy Sherif
Professor, University of Alberta, Edmonton, AB, Canada

Emico Okuno • Luciano Fratin

Biomechanics of the Human Body

Springer

Emico Okuno
Instituto de Física
Universidade de São Paulo
São Paulo, Brazil

Luciano Fratin
Faculdade de Engenharia
Fundação Armando Álvares Penteado
São Paulo, Brazil

ISSN 2192-4791   ISSN 2192-4805 (electronic)
ISBN 978-1-4614-8575-9   ISBN 978-1-4614-8576-6 (eBook)
DOI 10.1007/978-1-4614-8576-6
Springer New York Heidelberg Dordrecht London

Library of Congress Control Number: 2013947690

Translation of Desvendando a Física do Corpo Humano: Biomecânica, originally published in Portuguese by Editora Manole

Printed on acid-free paper

Springer is part of Springer Science+Business Media (www.springer.com)

Preface to the Edition in Portuguese

The idea of writing this book has been long-standing. The first real opportunity came in 1999, when our yoga teacher, Marcos Rojo, invited us to give a few physics classes to students of the Specialization Course in Yoga. Students came from different areas, such as journalism, medicine, architecture, and physical education. We prepared ten classes with the intention of teaching a topic of physics in each class and wrote class notes so that students could follow the lessons. At each class, we presented a theoretical part which was accompanied by a practical activity carried out with simple material in class. These classes served to calibrate the level and language being used: scientifically correct, without jargon, and easy to understand for non-physicists. To transform the notes into a book was a very long process.

We started from the premise that the human body is a laboratory that we have and that follows us wherever we go. So each person can test the concepts of classical mechanics, discussed here, in his or her own body. Furthermore, we have chosen to present the concepts in an objective manner and with a mathematical formalism suitable to a reader who has a high school education.

This book is designed to be used by undergraduate students in physical therapy, physical education, and other courses for those who have biomechanics in their curricula. High school teachers can also use this book as an alternative or as complementary material.

The mathematical language used in physics has always been blamed for the difficulties that students encounter in their studies. Therefore, in this book, every concept presented is followed by illustrative examples and in the sequence applications and exercises are proposed.

The content of the book is organized into eight chapters to allow a conceptual evolution of the student. Let's start with the last one, Chap. 8, of practical applications with simple experiments related to the concepts covered in each chapter that can be performed with materials easily found. It should not be left to the end, but rather we recommend its use at the end of or concurrent to each topic covered. Chapter 1 defines force, a vector quantity, and procedures necessary to carry out fundamental operations with this quantity and presents some specific types of forces. Chapter 2 introduces the concept of torque that, unlike the concept of force, is usually new to the student. This concept allows establishing the equilibrium conditions of a body. In this chapter, the student begins to envision new concepts, which he did not learn in high school. The torque of the weight force is discussed in Chap. 3, for the study of center of gravity, and how to determine it and the study of the stability of the human body, important for yoga practitioners, dancers, and practitioners of various sports. Chapter 4 complements the study of rotations introducing new concepts such as moment of inertia, radius of gyration, and angular momentum. Here the reader comes to understand numerous maneuvers performed by gymnasts and acrobats. Simple machines have been specially treated for physical therapists in Chap. 5: levers and pulleys used in treatments. In Chap. 6, the muscle forces associated with pain in the spinal column due to incorrect postures are determined, showing how to reduce their intensity quantitatively with the correct posture. Chapter 7 discusses the elastic properties of bone, stress and strain, pressure on the vertebrae, and broken bones in collisions.

Thus, we believe that this book will be useful and help to unravel the physics of the human body which is important for physical therapists, practitioners of sport in general, students of life sciences in a more broad sense, and non-physicists, who are curious and interested in learning. We would also recommend that high school teachers use examples from the book in order to motivate students to realize the importance of physics and start to like it and, who knows, love it.

Acknowledgments

We would like to thank the many colleagues who encouraged us and the students who tested our course. We could not fail to mention our dear yoga teacher, Professor Marcos Rojo, responsible for the groundbreaking of the process and Professor John Cameron of the University of Wisconsin, who, from afar, gave much support through e-mail.

We are also grateful to Daniela Pimentel Mendes and the team at Editora Manole for their commitment and dedication in editing the book in Portuguese version with utmost perfection. Finally, we thank our family for the love and understanding.

São Paulo, Brazil   Emico Okuno
São Paulo, Brazil   Luciano Fratin

Preface to the Edition in English

This book is a translation from the Portuguese of "Desvendando a Física do Corpo Humano: Biomecânica" published by Editora Manole Ltda. The publication of this English version is due to the patience, understanding, and efforts of Christopher T. Coughlin, physics publishing editor at Springer, and Fan, HoYing to whom we extend our sincere gratitude.

We are especially indebted to Renato Fratin who redid all of the drawings neatly for this edition.

We are also grateful to Prof. Eduardo Yukihara (Oklahoma State University) who made the first corrections of the English translation. Our deep gratitude goes to Prof. Wayne Seale (Instituto de Física/USP) for the comments, suggestions, revisions, and considerable improvements in the English translation, transforming brilliantly our Brazilian English constructions into proper English.

São Paulo, Brazil   Emico Okuno
São Paulo, Brazil   Luciano Fratin

Contents

Chapter 1
Forces

When muscles of the human body exert forces, they can set an object into motion or even change its state of motion. These muscular forces can cause deformation of bodies that is generally not visible to the unaided eye. Forces of many types control all motion in the universe.

1.1 Objectives

- To describe a force
- To represent a force by a vector using a scale and specifying its magnitude and direction
- To perform operations with vector forces and to determine the resultant force
- To calculate the pressure produced by a force on a surface

1.2 Concept of Force

Force is associated with a push (compression) or a pull (tension or traction) as shown in Fig. 1.1. Forces can produce motion, stop motion, or modify the motion of bodies on which they act. Forces can also deform the body on which they act. Forces are always applied by one body on another body.

A push on an object (e.g., a toy) uses a muscular effort to produce a movement that has the direction of this push. A pull on the toy in the opposite direction will reverse the motion. A force can be represented by a vector. The length of the vector, represented by an arrow, gives the magnitude of the force, and its tip indicates the direction. Force is measured in newtons (N) in the International System of Units (SI). Remember that magnitude and direction characterize a vectorial quantity. The forces in Fig. 1.1 are called contact forces, since these forces occur with two bodies in contact. The forces exerted by gases on the walls of a container, or our feet on the ground, are examples of contact forces.

E. Okuno and L. Fratin, *Biomechanics of the Human Body*, Undergraduate Lecture Notes in Physics, DOI 10.1007/978-1-4614-8576-6_1,

Fig. 1.1 (**a**) A small
horizontal force pushes an
object to the right. (**b**) A
larger force pulls an object
at an angle of 45° with the
horizontal to the right. Both
forces are applied, for
example, by the hand of a
person

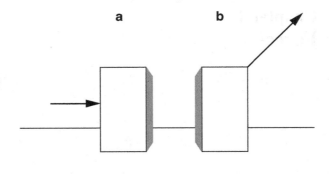

Fig. 1.2 Two examples of
forces acting at a distance.
A magnet attracts a piece of
iron, and a large mass M and
a small mass m experience a
gravitational attractive
force, in a different way
compared with the contact
forces shown in Fig. 1.1

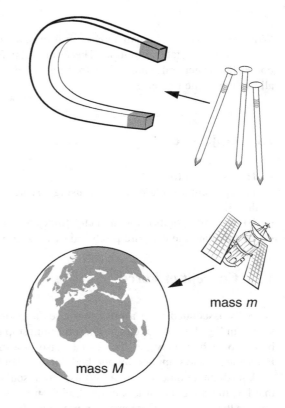

We also must consider forces that act at a distance, such as electric, magnetic,
and gravitational forces, shown in Fig. 1.2. In these cases, the source of the force is
not in contact with the body on which it acts, and the force is called a field force.
In this chapter, we deal with three types of force: gravitational force, muscle force,
and friction. The actions of gravitational and muscle forces cause joint compression
and joint tension, compression, or pressure (force per unit area) on the tissues or
organs of a body.

Exercise 1.1 Research and describe the laws of force of the interaction between electric charges (Coulomb's law) and gravitational attraction between bodies (Newton's universal law of gravitation). Specify the properties which give origin to such forces. Discuss the relationship between both forces and the distance between bodies. Discuss why, in the first case, forces can be either attractive or repulsive and, in the second case, there is only the force of attraction.

1.3 Representation of Forces: Diagram of Forces

Vectors are characterized by both magnitude and direction and can be represented graphically or mathematically. Force is an example of a vector quantity, and it is indicated by $\vec{F}$ or by boldface letter F. In a diagram, a vector is represented by an arrow whose direction determines the line of action; its length obeys a scale and is proportional to the magnitude or intensity of the force. The head of the arrow determines the direction of the vector and its origin, the location where the force is applied. In Fig. 1.3, three force vectors with different magnitudes (length = intensity) and directions are depicted. The magnitude of vector F is written F.

A system of coordinates can be used to represent a force vector. In the case of rectangular coordinates, shown in Fig. 1.4, a force can be described through its projection on each axis. The sign of a rectangular component is positive (+) or negative (−) when the arrow is directed upward and to the right or downward and to the left, respectively. The following trigonometric relations can be used:

$$\tan\theta = F_y/F_x, \ \sin\theta = F_y/F \text{ and } \cos\theta = F_x/F.$$

The magnitude F can be obtained by the Pythagorean theorem: $F = \sqrt{F_x^2 + F_y^2}$.

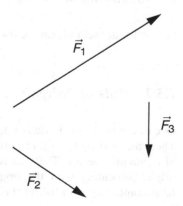

Fig. 1.3 Three force vectors with different magnitude and direction

Fig. 1.4 Vector force **F**
represented by its
components **F**$_x$ and **F**$_y$

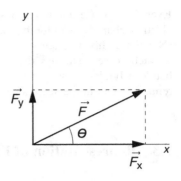

1.4   Resultant or Sum of Force Vectors

When two or more forces act on a body, it is possible to determine a force called resultant force, which can produce the same effect as all forces acting together. For this, we must know how to work with vector quantities. Some important rules are:

- There is the opposite vector: $-\vec{F}$ is the opposite vector of $\vec{F}$ with the same magnitude (intensity or length) but opposite direction.
- The multiplication of a vector $\vec{F}$ by a real number n is another vector $\vec{T}, \vec{T} = n \times \vec{F}$, with magnitude $T = nF$, same direction as $\vec{F}$, depending on the sign of n; that is, if n is positive, $\vec{T}$ will have the same direction as $\vec{F}$, and the opposite direction, if n is negative.
- The associative property is valid: $(\vec{F}_1 + \vec{F}_2) + \vec{F}_3 = \vec{F}_1 + (\vec{F}_2 + \vec{F}_3)$.
- The commutative property is valid: $\vec{F}_1 + \vec{F}_2 = \vec{F}_2 + \vec{F}_1$.
- A vector can be projected in a determined direction by using sine and cosine relations of a right triangle.

1.5   Addition of Vectors

Four rules or methods can be used to add vectors.

1.5.1   Rule of Polygon

One of the vectors is initially transported, maintaining its magnitude and direction. Then, the next vector is transported in a way that its origin coincides with the head of a previous vector. The sum vector or resultant vector will be an arrow with its origin coinciding with the origin of the first transported vector and with the head coinciding with the head of the last vector considered, as shown in Fig. 1.5.

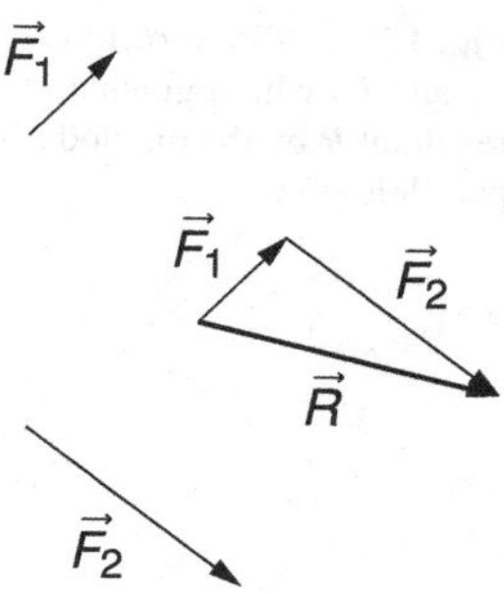

Fig. 1.5 Addition of
vectors F_1 and F_2, which
gives the resultant R by the
method of polygon

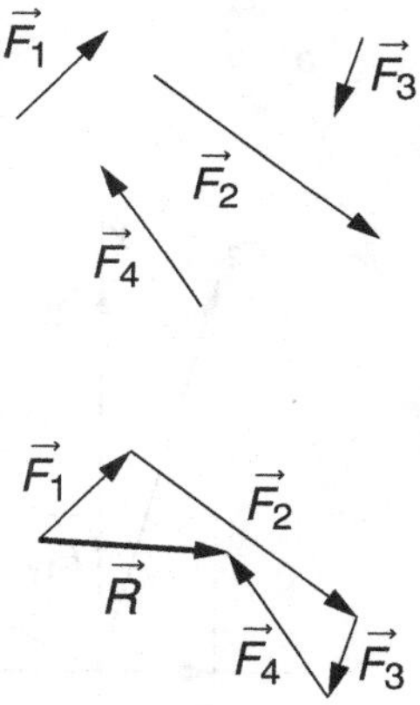

Fig. 1.6 Addition of
vectors F_1, F_2, F_3, and F_4
which gives the resultant R
by the method of polygon. It
is worthwhile to note that
the magnitude of the
resultant in this case is
smaller than that of Fig. 1.5

The magnitude of the sum vector can be obtained graphically, considering the scale adopted. This method can be applied to add any number of vectors by simply continuing this procedure; that is, the origin of the next vector should coincide with the head of the previous vector (see Fig. 1.6).

1.5.2   Rule of Parallelogram

Initially, both vectors are transported, maintaining their magnitude and direction, with their origins at the same point. Then, from the head of each vector, parallel lines to other vectors are drawn to form a parallelogram. The sum vector will be the arrow with the origin coinciding with the origin of vectors and the head, where the parallel lines cross, as illustrated in Fig. 1.7.

1.5.3   Method of Components

In this case, the vectors are represented in a system of rectangular coordinates and described as a sum of components (projections) in the x and y directions. The

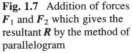

Fig. 1.7 Addition of forces F_1 and F_2 which gives the resultant R by the method of parallelogram

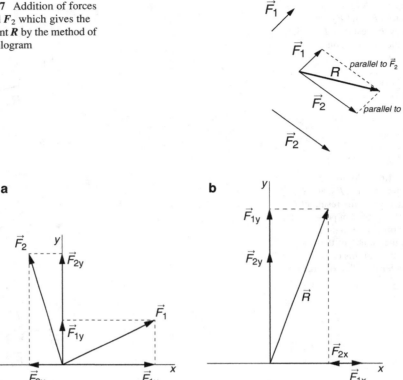

Fig. 1.8 Addition of vectors F_1 and F_2 decomposed in F_{1x}, F_{1y} and F_{2x}, F_{2y}, respectively, by the method of components in (**a**). In (**b**) the algebraic sum of F_{1x} with F_{2x} and F_{1y} with F_{2y} was done in order to obtain the resultant R

resultant vector obtained with the sum of several vectors will correspond to a vector in which its x (y) component is the algebraic sum of x (y) components of all vectors. Once the components of the vector are found, the magnitude of the resultant vector can be obtained by applying the Pythagorean theorem. This method is shown in Fig. 1.8, where the forces F_1 and F_2 are added.

1.5.4 Algebraic Method

The magnitude of the sum vector can also be calculated by the law of cosines, applied to the triangle formed by the forces F_1, F_2, and R, represented in Fig. 1.9:

$$R = \sqrt{F_1^2 + F_2^2 + 2F_1F_2 \cos\theta}. \qquad (1.1)$$

Fig. 1.9 Addition of
vectors with application
of the law of cosines

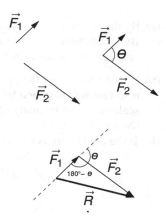

F_1 and F_2 are the magnitudes of the forces F_1 and F_2, respectively, and θ is the angle between the forces F_1 and F_2.

Example 1.1 The figure of Example 1.1 shows a way to exert a force on a leg of a patient in a traction device. The two tensions have the same magnitude $F_1 =$ weight W of the hanging object $= 45$ N. Obtain the net force (vector sum) acting on the leg:

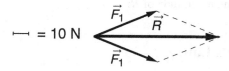

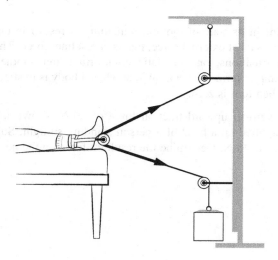

(a) By the method of parallelogram
(b) By calculation, applying the laws of trigonometry, considering that the angle between the forces is 50°

(a) We begin to solve the problem by adopting a scale as shown in the figure. The resultant was obtained by the method of parallelogram. Applying the factor of scale in R, the intensity of R is calculated as being 80 N. The direction of R is shown in the figure.
(b) If the angle between the forces is 50°:

$$R = (45^2 + 45^2 + 2 \times 45 \times 45 \times \cos 50°)^{1/2} = 82\,\text{N}.$$

$$R = 82\,\text{N}.$$

Note: The precision of the graphic method is not very good. It can be improved by drawing the figure in a larger scale.

Exercise 1.2 The leg of Example 1.1 is now moved away such that the angle between the forces will be 30°, maintaining the same value of force $F_1 = W$:

(a) In this case, will the value of the magnitude of R be greater or smaller than the answer of Example 1.1?
(b) Determine the magnitude of R.

1.6 Newton's Laws

1.6.1 Newton's First Law of Motion (Law of Inertia)

A body will maintain its state of motion, remaining at rest or in uniform motion unless it experiences a net external force, that is, a resultant force. This law implies two equilibrium situations, one of static equilibrium and another of dynamic equilibrium. In simpler terms, we can say that when a body is in static equilibrium, the net force applied to it is zero.

Exercise 1.3 A vertical upward traction force of 60 N, shown in the figure of Exercise 1.3, is applied on a head of a person standing straight. Suppose that the weight of the head is 45 N. Determine the resultant force on the head.

1.6.2 Newton's Second Law (Mass and Acceleration)

The action of a nonzero resultant force on a body produces change in the vector velocity, that is, produces an acceleration a. This acceleration is proportional to the intensity of a net force F and inversely proportional to the mass m of the body, that is,

$$a = F/m.$$

Then, we can write that $F = ma$. The unit of velocity in SI is m/s, and as acceleration is given by

$$a = \Delta v/\Delta t,$$

that is, the rate of change of speed Δv with time Δt, its unit in SI is m/s^2. Therefore, the unit of force is kg m/s^2 which receives the special name newton, N, to honor the father of classical mechanics, Isaac Newton (1642–1727).

Exercise 1.4 What force should act on a ball of 0.6 kg mass, through a kick, to acquire an acceleration of 40 m/s^2?

1.6.3 Newton's Third Law (Action and Reaction)

Force is a consequence of the interaction between two bodies. The third law states that for each action force corresponds a reaction force of equal intensity but opposite direction. Action and reaction act on different bodies. In the examples of Fig. 1.1, the forces (action forces) applied on an object were depicted. The forces of reaction were not drawn. The reaction forces are forces applied by the object on the hand that pulls or pushes the object.

Exercise 1.5 In high jumping, an athlete exerts a force of $3,000 \text{ N} = 3 \times 10^3 \text{ N}$[1] $= 3 \text{ kN}$ against the ground during impulsion. Find the force exerted by the ground on the athlete. Discuss the nature of this force.

1.7   Some Specific Forces

1.7.1   Weight

The weight of a body or of any object is the force with which a body is attracted to the earth. This explains the fact that an object always falls when released at a certain height from the ground. This force, also called gravitational force, or simply weight, is exerted by the earth on bodies not necessarily in contact. Other examples of force among bodies that do not need to be in contact are electric forces and magnetic forces (force exerted by a magnet on metallic objects, see Fig. 1.2). Gravitational force is always directed toward the center of the earth, as illustrated in Fig. 1.10. The magnitude of weight vector W is then given by (1.2):

$$F(\text{N}) = W(\text{N}) = mg. \tag{1.2}$$

m is the mass of the body, measured in kilograms (kg), and g is the acceleration of gravity, equal to 9.8 m/s^2 in any location near the surface of the earth. As we move toward outer space, away from our planet, the value of g decreases, which can be observed in spaceship travels (weightlessness). In this book, we will use $g = 10 \text{ m/s}^2$. The weight of a body with mass equal to 1.0 kg, where the acceleration of gravity is 9.8 m/s^2, is 9.8 N, practically equal to 10.0 N. Using this g value, we are making an approximation of 2 % in excess.

The weight of the human body on the moon is approximately 1/6 that on the earth, although the mass is the same, due to the low value of the acceleration of gravity on the moon, 1.6 m/s^2. The astronauts who walked on the moon felt lighter, released from the weight of their bodies and while walking looked as if they were hopping. If their total mass, including what they were wearing, was 100 kg, their weight on earth would be around 1,000 N, while on the moon, it would decrease to 160 N.

The reaction to the weight W (action), exerted by the earth on a body, is the force that the body exerts on the earth (reaction force R) and acts on the center of the earth. Its magnitude is the same as the weight and is in the opposite direction.

---

[1] Scientific notation: very large or very small numbers can be written, using powers of 10: $175,000 = 1.75 \times 10^5$ or $0.000175 = 1.75 \times 10^{-4}$.

Fig. 1.10 Two bodies of different masses on the surface of the earth, therefore with different weights and respective reaction forces

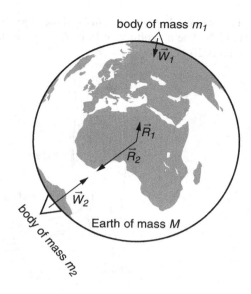

Figure 1.10 shows the weights of two bodies with different mass and respective reaction forces exerted by bodies on the earth.

1.7.2 Muscle Forces

Animal posture and motion are controlled by forces produced by muscles. There are approximately 600 muscles in the human body that are responsible for all the motions of the body from very subtle movements in the facial expression to moving the tongue in speech, to circulating the blood in the vessels of the body, and to the beating of the heart, whose main function is muscle contraction.

A muscle consists of a large number of fibers whose cells are able to contract when stimulated by nerve impulses coming from the brain. A muscle is usually attached to two different bones by tendons.

The maximum force that a muscle can exert depends on its cross-sectional area (perpendicular cut) and is inherent to the structure of muscle filaments. This maximum force per unit area varies from 30 to 40 N/cm^2. It does not depend on the size of the animal and, therefore, has the same value for a muscle of a rat or of an elephant. Under the microscope, the muscle of an elephant is very similar to that of a rat, except in the quantity of mitochondria, which is larger in smaller animals.

Example 1.2 Representation of motionless (at rest) biceps muscle, bones of arm, and forearm with an object in the hand can be seen in the figure of Example 1.2. The forces that act on both forearm and hand are drawn in the figure of Example 1.2. Find the magnitude of force exerted by the biceps muscle, by adding all the forces, observing that once the system is in equilibrium, the resultant must be equal to zero.

The magnitude of the weight **W** of the object plus hand is 20 N and of the forearm A is 15 N. The magnitude of force **R** which is the reaction of the humerus against the ulna is 20 N.

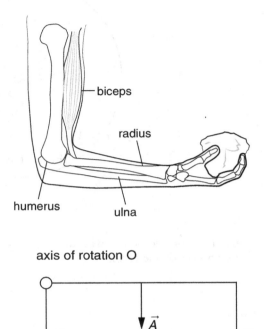

Transporting all the forces to the same line, we see that there are three downward forces totaling 55 N. Therefore, to equilibrate this resultant, the muscle force exerted by the biceps must have an intensity of 55 N but with opposite direction, that is, directed upward.

1.7.3   Contact Force or of Reaction or Normal (Perpendicular) Force

Consider a block at rest on a table, as shown in Fig. 1.11. Let us see what forces act on the object. The block experiences a force **W** due to the gravitational pull of the earth. As the block is at rest, the resultant of all applied forces must be zero. Therefore, another force of equal magnitude and opposite direction is applied by the surface of the table. This is the contact force or normal force **N**, meaning perpendicular to the surface. The reaction (not depicted) to the force **W** is exerted on the earth, and the reaction to **N** is another contact force $N^* = -N$, exerted by the block on the surface of the table.

Fig. 1.11 Two forces are applied on the block: the force of gravity (weight) W and the normal force N whose sum gives zero resultant, as the block is at rest. The normal force N^* is the reaction force to the action force N. The reaction to W is applied on the center of the earth and is not represented here

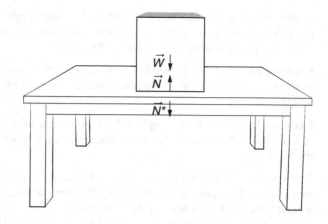

Example 1.3 Consider two blocks: block A, which weighs 5 N, is on the table. Block B, weighing 10 N, is on top of block A as shown in the figure of Example 1.3. Analyze the forces that act on each block separately. Finally, determine the total contact force exerted by the two blocks on the table.

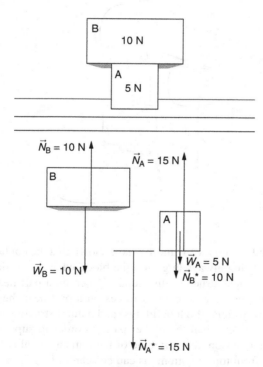

As the two blocks are at rest, the resultant of applied forces on each block must be zero. The weight of the table was not considered, but only the contact force that

results from the two blocks on the table. The magnitude and the direction of the forces are shown in the figure. Note that the table should support the contact force of 15 N, due to the weight of the two blocks on it.

Exercise 1.6 Consider Example 1.3. Now invert the position of the two blocks and analyze the forces. Draw all the forces that act on each block and the contact force on the table.

Exercise 1.7 Consider Example 1.3. Now place a third block which weighs 3 N beneath block A. Draw all the forces that act on each block and the contact force on the table.

Exercise 1.8 Consider a person seated with the leg vertically and the foot hanging, as shown in the figure of Exercise 1.8. This person wears an exercise boot with a weight attached to it. Consider that the sum of the mass of the leg with that of the foot is 3.5 kg (35 N), that the mass of the boot is 1.35 kg (13.5 N), and that the mass of the weight is 4.5 kg (45 N). Represent the forces that act on the leg and determine the resultant of forces which must be compensated by the force acting on the knee articulation (joint).

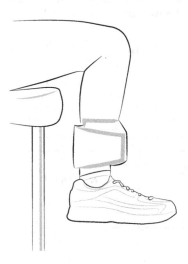

   Example 1.3 and Exercises 1.6 and 1.7 have shown that the contact force on each block is different, being always larger on the block which is on the bottom of the structure. In any vertical structure, the contact force on a part near the bottom of the structure is greater than the contact forces on a part near the top. This is the reason why in both artificial (high buildings) and natural structures (spinal column) the lower parts are larger than the higher parts in order to support larger contact forces. For the same reason, the vertebrae of the human spinal column increase in size continuously from top to bottom, as can be seen in Fig. 7.6.

Fig. 1.12 Force of friction
as a function of applied
force. The values of
maximum force of static
friction and the force of
kinetic friction are shown in
the figure

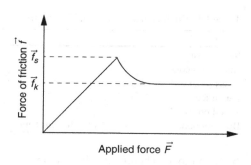

Example 1.4 Consider a man with a mass of 70 kg. The mass of the head plus the neck is 5.0 kg. Find the intensity of the normal force (of contact) exerted mainly by the seventh cervical vertebra which supports the head and the neck.

As the weight of the head plus the neck is 50 N and the body is at rest, the answer to this question is also 50 N.

Exercise 1.9 The mass distribution of the body of a man with 70 kg is the following: head plus neck (5.0 kg), each arm-forearm-hand (3.5 kg), torso (37 kg), each thigh (6.5 kg), and each leg plus foot (4.0 kg). Supposing that this person is standing upright on both feet, find the intensity of the normal force (of contact): (a) exerted at each of the hip joints and (b) exerted at each of the knee joints. Supposing now that the man stands on one foot, find the intensity of the contact force: (c) in the knee joint of the leg by which the man is supported and (d) in the knee joint that supports the leg that is off the floor.

1.7.4   Forces of Friction

The force of friction f is a force with which a surface in contact with the body applies on it, when submitted to a force, in the same way as the contact force, with the fundamental difference that the contact force is always perpendicular to the surface and the force of friction is parallel to the surface. Unlike all the forces previously discussed, that of friction appears in bodies in motion or on the verge of moving. It has an opposing direction to that of the external applied force, and hence, it opposes the movement. The origin of this force is in the roughness of both surfaces in contact.

If the applied force F on a body at rest is not enough to move it, that means that there is a frictional force f of equal magnitude and opposite direction in a way that the net force is zero, since $F = -f$. Therefore, as the intensity of the applied force is increased, the intensity of the force of friction follows this increase in the same way as shown in Fig. 1.12 and the body remains stationary. However, as the intensity of the frictional force can be increased up to a maximum value called maximum force of static friction f_s, an applied force with magnitude above this value will cause the

Table 1.1 Coefficients of static and kinetic friction

Materials	μ_s	μ_k
Steel on steel	0.74	0.57
Rubber on concrete	1.00	0.80
Glass on glass	0.94	0.40
Ice on ice	0.10	0.03
Wood on wood	0.25–0.50	0.20
Bone on bone with synovial fluid in human beings	0.01	0.003

body to move. Therefore, when an applied force overcomes f_s, the body begins to move. Experimentally, it is found that

$$f_s = \mu_s N, \tag{1.3}$$

where μ_s is the coefficient of static friction and N is the normal force (1.3), whose intensity is equal to that of the weight of the body.

When the body is in motion, a smaller applied force is enough to maintain a constant velocity. This force is called force of kinetic friction f_k and can be obtained from (1.4):

$$f_k = \mu_k N, \tag{1.4}$$

where μ_k is the coefficient of kinetic friction.

The values of μ_s and μ_k depend on the nature of the surfaces in contact but are almost independent of the surface areas in contact. They are dimensionless, that is, net numbers without units.

Table 1.1 lists the values of coefficients of static and kinetic friction between different materials. It is possible to observe from the table that the synovial fluid causes the coefficient of friction in bone joints to have a much smaller value than the coefficients of friction between other materials. This fluid acts as a lubricant in order to facilitate the movement after the frictional force is overcome. The lubricants used, for example, in the motors of cars have exactly the same role, that is, to facilitate the motion, besides reducing the erosion of materials. In the human body, there are many fluids that have this function. Saliva acts as a lubricant making the deglutition (swallowing) of food possible. Its lack would make it painful for the swallowing of toasts and granola, for example.

However, the existence of friction is essential in many situations. When we walk or run, as the heel of the foot touches the ground, the foot pushes it in a forward direction and the ground, in its turn, exerts a frictional force in the backward direction, preventing the person from slipping. When the toe leaves the ground, the frictional force prevents the toe from slipping backward. Therefore, we would not be able to walk or run on a surface without friction. Who has not fallen down while walking or running on wet and slippery or icy ground or on soapy or well-waxed floors? Going for a walk on smooth ground, it is important to wear tennis shoes with rough soles to increase the friction. The rubber tires of vehicles rotate without effect and slip when there is oil or ice on the ground, the cause of many accidents due to the lack of friction.

Example 1.5 I decided to move some furniture. I have begun pushing a file cabinet full of papers, with a mass of 100 kg. For this, I applied a force of 200 N, but the file cabinet remained in its place. I had to ask a friend for help. Together we could double the force to 400 N. Consider that the coefficient of static friction between the table and the ground is 0.5 and the coefficient of kinetic friction is 0.3.

(a) Find the force of static friction that has acted on the file when I applied force equal to 200 N.
(b) Evaluate if we had success in pushing the file cabinet with the help of a friend. Justify your answer.
(c) Determine the intensity of force that must be applied to put the file cabinet in motion.
(d) Verify if it was possible to dispense the help of friends, after the file cabinet was in motion.

(a) $f = 200$ N, since the file cabinet remained at rest.
(b) $f_s = \mu_s N = 0.5(100 \text{ kg})(10 \text{ m/s}^2) = 500$ N. This is the minimum force that must be applied on the file cabinet to begin moving it. Therefore, the effort of two persons was not enough and needed the help of a third person.
(c) >500 N.
(d) $f_k = \mu_k N = 0.3(100 \text{ kg})(10 \text{ m/s}^2) = 300$ N. Therefore, once in motion, it was possible to dispense the help of the third friend, but not of the second.

Exercise 1.10 The static friction between a tennis shoe and the floor of a basketball court is 0.56, and the normal force which acts on the shoe is 350 N. Determine the horizontal force needed to cause slippage of the shoe.

Example 1.6 Consider a child with 20 kg mass playing on a slide that makes an angle with the horizontal of 45°, as illustrated in the figure of Example 1.6. The coefficients of static and kinetic friction between the body of the child and the slide are 0.8 and 0.6, respectively.

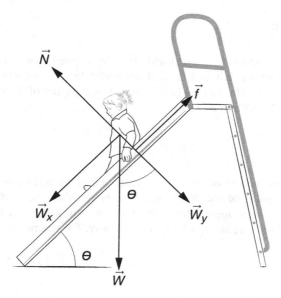

(a) Decompose the weight W of the child in orthogonal components W_x and W_y in relation to the plane of the slide and obtain the value of these components.

(b) Find the value of the normal force exerted by the surface of the slide on the child.

(c) Evaluate if the child will slide down when she or he lets go.

(d) Find the force of kinetic friction.

(e) Determine the acceleration of the child sliding down.

(f) Discuss what happens if the angle θ is greater than 45°.

(a) If the angle θ between the slide and the ground is 45°, the angle between W_y and W will also be 45°, since both sides of the angles are mutually perpendicular. Then,

$$W_y = W\cos\theta = (20\,\text{kg})(10\,\text{m/s}^2)\cos 45° = (200\,\text{N})0.707 = 141.4\,\text{N}.$$

$$W_x = W\sin\theta = (20\,\text{kg})(10\,\text{m/s}^2)\sin 45° = (200\,\text{N})0.707 = 141.4\,\text{N}.$$

(b) As $N = W_y$, $N = 141.4$ N.

(c) $f_s = 0.8\,N = 0.8(141.4\,\text{N}) = 113.1$ N; the child goes down the slide, since W_x is greater than f_s.

(d) $f_k = 0.6\,N = 0.6(141.4\,\text{N}) = 84.8$ N.

(e) $a = (W_x - f_k)/m = (141.4 - 84.8\,\text{N})/(20\,\text{kg}) = 2.83\,\text{m/s}^2$.

(f) If the angle θ is increased, for example, to 60°, $\cos 60° = 0.5$ and $\sin 60° = 0.866$. In these conditions, $W_y = (200\,\text{N})0.5 = 100$ N decreases, and, as a consequence, the value of normal force N decreases. Then, $f_s = 0.8 (100\,\text{N}) = 80$ N will also be smaller, meaning that the child will slide more easily and the acceleration will be greater. If, on the other hand, the angle is decreased, the opposite happens.

1.8   Pressure

The concept of pressure is associated with the force applied on a body. Pressure p is defined as the force per unit area exerted perpendicularly on a surface.

Pressure p on the surface of the block exerted by the palm of the hand, as shown in Fig. 1.13, can be written by (1.5):

$$p = \frac{F_y}{A}, \tag{1.5}$$

where F_y is the component of force F perpendicular to the surface of block and A is the area of the palm of the hand. Pressure is inversely proportional to area. Therefore, for the same applied force, the pressure will be greater the smaller the area. Knives and scissors also work in this same way. The sharper they are, the more

Fig. 1.13 A force F is applied by the palm of a hand with area A on the surface of a block. This force has been decomposed in a horizontal component and a vertical component, perpendicular to the surface of the block

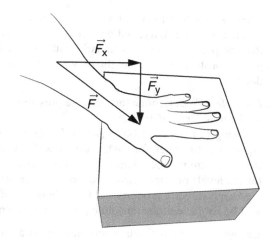

easily they cut. A nail has a pointed extremity, just to facilitate penetration, and causes great pressure.

The unit of pressure in SI units is N/m^2 with the special name pascal (Pa). Its plural is pascals, as it comes from the name of the famous scientist Blaise Pascal (1632–1662). For fluids, other units are used. Blood pressure, for example, is measured in millimeters of mercury (mmHg) and the pressure of the eyeball, of the bladder, etc., in centimeters of water (cmH$_2$O). Normal blood pressure of an adult is between 80 and 120 mmHg. Normal voiding pressure of the bladder is around 30 cmH$_2$O.

In the case of gases, pressures are measured in atmospheres. Atmospheric pressure is the pressure exerted by the atmosphere which is made up of atoms and molecules of air, on the earth's surface. Its value at sea level is 1 atmosphere (1 atm). The higher the altitude, the smaller is the quantity of atmosphere above it. For this reason, the pressure in such locales (high altitude) is smaller than at sea level. In Chacaltaya, Bolivia, at 5,000 m of altitude, the atmospheric pressure is around 0.5 atm, where many persons not used to it feel the lack of oxygen. Today, when the weather forecaster announces the atmospheric pressure in the coastal cities, they say that it is 1,013 HPa (hectopascal) which is equal to 1.013×10^5 Pa.

The pressure exerted by compressed air in a tire is measured in psi (pounds per square inch). So, the tires are calibrated with 30 psi, for example.

The relationships among different pressure units are:

1 atm = 760 mmHg.
1 atm = 1.03×10^3 cmH$_2$O.
1 atm = 1.013×10^5 Pa.
1 atm = 14.7 psi.

The pressure p exerted by a water column of height h can be calculated by using (1.6):

$$p = \rho g h, \qquad (1.6)$$

where ρ is the density[2] of water $= 1$ g/cm$^3 = 1,000$ kg/m$^3 = 10^3$ kg/m^3, g the acceleration of gravity, and h the height of the water column. Equation (1.6) shows that the pressure increases linearly with the height of the water column. As ρ and g are constants, if the height of water column is doubled, the pressure is also doubled.

Example 1.7 Determine the absolute pressure on the body of a person who dives in a lake to a depth of 10 m.

Using (1.6), $p = (1,000$ kg/m$^3)(10$ m/s$^2)(10$ m$) = 10^5$ Pa which is practically equal to 1 atm. Therefore, when diving and a depth of 10 m is reached, the absolute total pressure will be 2 atm, which is the sum of the atmospheric pressure of 1 atm (at sea level) plus the pressure exerted by the 10 m of the water column.

As p is directly proportional to the height h, if the diver reaches 20 m, he or she will be subjected to an absolute pressure of 3 atm.

Example 1.8 Two children are playing on a seesaw, whose arm can incline at a maximum of 30° relative to the horizontal. The mass of one child is 20 kg and the other is 21 kg. They are playing well, giving small push offs from the ground. At a given moment, the child with the smaller mass was up and at rest. Find the pressure exerted by this child on the board, remembering that her or his contact area with the board is 300 cm$^2 = 0.03$ m^2.

The weight W of the child is $(20$ kg$)(10$ m/s$^2) = 200$ N. The gravitational force is always perpendicular to the ground. Therefore, we have to find the normal component W_y of the gravitational force on the plane of the seesaw:

$$W_y = W \cos 30° = (200\,\text{N})0.866 = 173.2\,\text{N}.$$

Therefore, $p = 173.2$ N/0.03 m^2; $p = 5,773.3$ Pa.

Example 1.9 Find the pressure on the ground exerted by each foot of a child with mass of 20 kg, when she or he is standing on two feet. Consider the area of each foot as being 60 cm^2.

The weight to consider is $W = 100$ N, which is the half of the weight of the child, pressing the ground with the sole of each foot:

$$p = 100\,\text{N}/0.0060\,\text{m}^2; \quad p = 16,667\,\text{Pa}.$$

Note the increase in this pressure compared to that of the Example 1.8, since the area of contact was decreased.

Exercise 1.11 Consider now that the child of 20 kg is standing on one foot only. Find the pressure exerted by this foot on the ground. Estimate now the pressure on the ground in case the child stands on the toe of one foot, with a contact area of 8 cm^2?

---

[2] Density of a substance $\rho = m/V$, that is, the ratio between the mass m of substance and volume V which contains the mass m.

Examples 1.8 and 1.9 and Exercise 1.11 have shown clearly that the pressure exerted by the weight of a person will become larger as the contact area of this person with the ground becomes smaller. In this way, if the seat of a couch supports the pressure of the body when the person is seated, the couch cannot necessarily support the weight if this person stands on it, mainly on one foot. The effect of the area on the pressure is easily verified in the sand of a beach, by changing from a tennis shoe to a high-heeled shoe.

The concept of pressure will be reviewed and discussed in terms of compression on intervertebral discs in Chap. 7.

1.9 Answers to Exercises

Exercise 1.1 Electric and gravitational forces are, respectively, equal to $F_E = K\dfrac{q_1 q_2}{r^2}$ and $F_G = G\dfrac{m_1 m_2}{r^2}$. In the case of electric forces, they can be positive (repulsive) or negative (attractive), depending on the sign of the charges q_1 and q_2 with both positive or both negative and one of them positive and the other negative, respectively. It is known that charges of opposite signs attract and same signs repel. In the case of gravitational force, there is only the force of attraction, since there is no negative mass. The intensity of these forces in the first case depends on the values of the charges and in the second case on the masses of the bodies. The constants of proportionality $K = 9 \times 10^9$ N m^2/C^2 and $G = 6.67 \times 10^{-11}$ N m^2/kg^2 are universal constants. Note that, as G is very small, the gravitational attraction between any two bodies is very difficult to observe. Both forces decrease with the square of the distance.

Exercise 1.2 (a) Will be larger; (b) $R = 87$ N. Note that, since the weight, F_1, is maintained, the resultant of the applied force can be changed by modifying the angle between the forces, which can be done by changing the position of the leg horizontally.

Exercise 1.3 $R = 15$ N upward, pulling the head.

Exercise 1.4 $F = 24$ N.

Exercise 1.5 Force of reaction $= 3,000$ N $= 3 \times 10^3$ N. It is a contact force.

Exercise 1.6 The gravitational force of 5 N and the normal force of 5 N are applied on block A. The gravitational force of 10 N, normal force of 5 N exerted by block A, and the normal force of 15 N are applied on block B. The total force on the table due to the weights of the two blocks is 15 N. Observe that the value of the last force does not depend on the order of the blocks placed above it.

Exercise 1.7 The forces on blocks A and B are the same as in Example 1.3, without the normal due to the table. The weight of 3 N acts on block C, the normal force of 15 N exerted by block A and the normal force of 18 N. The total force on the table is 18 N. Observe that the force on the table is the sum of the weights of all of the blocks placed above it.

Exercise 1.8 $R = 35$ N + 13.5 N + 45 N = 93.5 N.

Exercise 1.9 (a) 245 N; (b) 310 N; (c) 660 N; (d) 40 N.

Exercise 1.10 F is greater than 196 N.

Exercise 1.11 (a) $p = 3.33 \times 10^4$ Pa; (b) $p = 2.50 \times 10^5$ Pa. The last value is around 2.5 atm.

Chapter 2
Torques

Torque exerted by a force is an important physical quantity in our daily life. It is associated with the rotation of a body to which a force is applied, unlike the force that is related to translation. For a body to be in rotational equilibrium, the sum of all torques exerted on it must be zero.

2.1 Objectives

- To discuss the concept of torque
- To obtain the torque due to more than one force
- To establish the conditions for rotational equilibrium of a rigid body

2.2 Concept of Torque

Torque or moment of a force, M_F, is a physical quantity associated with the tendency of a force to produce rotation about any axis.

Torque is a vector quantity, but, in this book, we will use it as a scalar, introducing a sign convention that will allow us to add algebraically several torques due to the forces applied on a body. The sign of torque is taken to be positive (+) if the force tends to produce counterclockwise rotation and negative (−) if the force tends to produce clockwise rotation about an axis.

The effect of rotation depends on the magnitude of the applied force F and on the distance $d_\perp$ (perpendicular) to the axis of rotation. Torque is calculated by the product of the magnitude of force by the distance $(d_\perp)$ from the line of action of force F to the axis of rotation. The line of action is the straight line, imaginary, that determines the direction of the force vector. The distance $d_\perp$ is called the moment arm or the lever arm of the force F. The segment that defines the lever arm is perpendicular to the line of action of the force and passes through the axis of rotation. The magnitude of the torque, M_F, is defined by (2.1):

E. Okuno and L. Fratin, *Biomechanics of the Human Body*, Undergraduate
Lecture Notes in Physics, DOI 10.1007/978-1-4614-8576-6_2,
© Springer Science+Business Media New York 2014

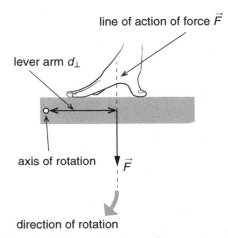

line of action of force $\vec{F}$

lever arm $d_\perp$

axis of rotation $\vec{F}$

direction of rotation

Fig. 2.1 A force F is applied at the center of an object. The line of action of the force F, the lever arm $d_\perp$ of the force F, and the axis of rotation through O are represented. The *arrow* indicates the direction of rotation

$$M_F = Fd_\perp. \tag{2.1}$$

Its unit in the International System of Units (SI) is N m.

Figure 2.1 shows a force F applied at the center of an object. The rotation axis is an imaginary line, perpendicular to the pivot point O, or fulcrum, of the object. The lever arm $d_\perp$ is perpendicular to the line of action of the force F, which tends to rotate the object about the pivot point O clockwise; the magnitude of torque is $M_F = -Fd_\perp$. Now, if the force F with the same magnitude is applied in the opposite direction at the same place, the direction of rotation changes; that is, the body will rotate counterclockwise and the torque is given by $M_F = Fd_\perp$. Note that the axis of rotation is always subjected to the reaction (not shown) to the force F. However, this reaction force does not produce torque because its lever arm is equal to zero. You can imagine that Fig. 2.1 represents a top view of a cross section of a door and the axis of rotation is the hinge line and the force is applied to open the door.

If a force F of the same magnitude is now applied at the opposite extremity to the axis, as can be seen in Fig. 2.2, its torque will be twice that of Fig. 2.1 because the lever arm will be $2d_\perp$, i.e., $M_F = -F(2d_\perp)$. In other words, it will now be twice as easy to rotate a bar about its axis of rotation. That is what we usually do instinctively to open a door or when we use a wrench to loosen a nut.

If we now apply a force F at the same place and with the same magnitude as that of Fig. 2.2, but with a different direction, as shown in Fig. 2.3, we see that the torque will be zero because the line of action of force will pass through the axis of rotation and, in this case, the lever arm will be null. If we apply a force to a door in such a way, we will never be able to either open or close it.

If we now apply a force F of the same intensity as that of Fig. 2.2, but with a different direction, as shown in Fig. 2.4, we see that the torque, i.e., the turning effect, is decreased, due to the fact that the lever arm which must be perpendicular

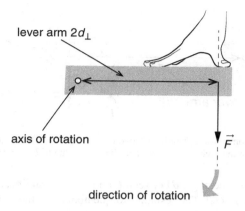

Fig. 2.2 A force **F** of the same intensity as that of Fig. 2.1 is now applied at the opposite extremity to the axis. The torque will be twice that of Fig. 2.1 because the lever arm was doubled

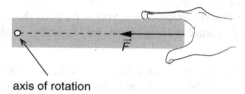

Fig. 2.3 A force **F** of the same magnitude as that of Fig. 2.2 is now applied at the end of the bar, but perpendicular, and the torque is null because the lever arm is zero, since the line of action, imaginary over the force vector, passes through the axis of rotation

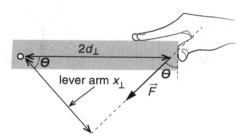

Fig. 2.4 A force **F** of equal intensity but different direction to that of Fig. 2.2 is applied at the edge of the bar. In this case, the lever arm $x_\perp$, which must be perpendicular to the line of action of force, is shorter than that of Fig. 2.2 and, as a consequence, the magnitude of the torque will be less

to the line of action of force **F** will be shorter, i.e., $x_\perp < 2d_\perp$. If we want to rotate the bar with the same ease, we have to apply a force with a larger magnitude. In the case of Fig. 2.4, the torque is $M_F = -Fx_\perp$.

We obtain the same result, decomposing the force **F** in its orthogonal components (Fig. 2.5), as presented in Chap. 1. The intensity of the component F_y can be

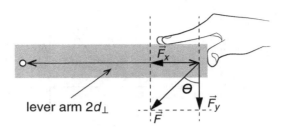

Fig. 2.5 Decomposition of force F in its orthogonal components. The torque due to the component F_x is zero because the lever arm is zero and the torque due to F_y is given by $F_y(2d_\perp)$

calculated if we know the angle θ between F and the vertical line, applying the law of trigonometry: $F_y = F\cos\theta$. The resultant torque about θ is then given by

$$M_F = F_x0 - F_y(2d_\perp) = -F\cos\theta\ (2d_\perp) = -Fx_\perp.$$

The torque due to the component F_x is zero and, therefore, can produce no rotation because the line of action of the vector F_x goes through the pivot point O. Figure 2.4 shows that since the angle between $2d_\perp$ and $x_\perp$ is also θ, we can write that $\cos\theta\ (2d_\perp) = x_\perp$.

Example 2.1 In the exercise of lateral lifting of an arm, one object with a 2 kg mass is held by the hand, as can be seen in the figure of Example 2.1. The length of the arm + forearm + center of the hand is 70 cm. The axis of rotation is located at the shoulder. Find the torque exerted by the weight of the object for each situation in which the angle between the arm and the body is (a) 30° downward and (b) 90°.

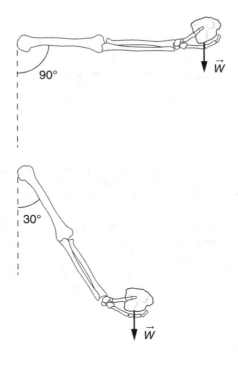

(a) $M_W = - mgd_\perp = -(2 \text{ kg})(10 \text{ m/s}^2)d_\perp$; the negative sign indicates that the rotation is clockwise around the shoulder.

$$\text{As } d_\perp = (0.70 \text{ m})\sin 30° = (0.70 \text{ m})0.5 = 0.35 \text{ m},$$

$$M_W = - (2 \text{ kg})(10 \text{ m/s}^2)(0.35 \text{ m}) = - 7 \text{ N m}.$$

(b) $M_W = - (2 \text{ kg})(10 \text{ m/s}^2)(0.7 \text{ m}) = - 14 \text{ N m}.$

To maintain the arm outstretched horizontally ($\theta = 90°$) in equilibrium, preventing its fall, it is necessary to apply a muscle force to counteract the torque. This force will be twice as large as the force to maintain the arm at 30°.

Let us see now what happens when the same object is hanging at the elbow with the arm outstretched horizontally. In this case, the torque will decrease to almost half of that with the weight on the hand, because the lever arm has been decreased to almost half. This means also that the muscle has to exert a force around half of the previous example to maintain it outstretched.

Exercise 2.1 Consider that a package of sugar of 1 kg is in the hand of a horizontally outstretched arm. Find the torque due to the weight of the sugar package about the axis of rotation passing through the:

(a) Wrist
(b) Elbow
(c) Shoulder
  Consider the following distances: shoulder–elbow = 25 cm, elbow–wrist = 22 cm, and wrist-center of the hand = 6 cm.
(d) Repeat the exercise, considering that the arm is held upward at a 30° angle to the body and then downward at the same angle.

Exercise 2.2 Discuss the reason for progressive difficulty in abdominal exercises, lying flat on the back: (a) with outstretched arms and the hands in the direction of the feet, (b) with the arms crossed over the chest, and (c) with the fingers interlaced under the head.

Exercise 2.3 Consider a person in the yoga posture called sarvanga-asana, as shown in the figure of Exercise 2.3. At the beginning, the person is lying flat on his back, with his legs outstretched. Then, slowly, the legs are lifted until the toes are pointed straight up. To reach the final posture, it is recommended that persons with problems in their spinal column bend their legs before lifting them. Discuss why.

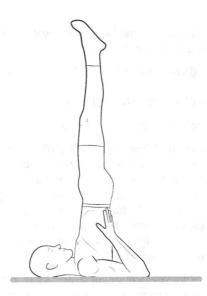

Fig. 2.6 Two forces are
applied on the bar. The net
torque is the sum of the
torques due to each force

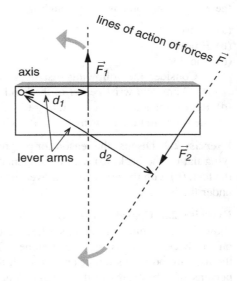

When a body is subjected to more than one force, we have to find the torque due to
each force and add them to obtain the resultant torque. Figure 2.6 shows two forces
acting on a bar. The resultant torque about the axis at pivot point O is given by

$$M = F_1 d_1 - F_2 d_2. \tag{2.2}$$

The direction of rotation is defined by the magnitude of forces and the lever arms
involved. If the first (second) term of (2.2) is smaller than the second (first) term, the
resultant torque will produce clockwise (counterclockwise) rotation of the bar.

2.3   Binary or Couple

Binary or couple is a system formed by two forces of the same magnitude and opposite direction applied on a body, whose lines of action are separated by a nonzero distance called the binary arm. The forces applied to the tap wrench or to a round doorknob constitute a couple. Figure 2.7 shows a sketch of a couple.

We can find the torque of a couple (2.3) by calculating the torque of each force about the axis of rotation separately and then summing them:

$$M = Fd_{\perp 1} + Fd_{\perp 2} = F(d_{\perp 1} + d_{\perp 2}) = Fx. \tag{2.3}$$

Both forces tend to rotate the bar counterclockwise about an axis through O.

Exercise 2.4 To remove the lug nut that fixes the wheel of a car, a man applies forces with a magnitude of 40 N with each hand on a tire iron, maintaining the hands 50 cm apart. Draw a diagram representing this situation and calculate the torque of the couple of forces exerted by the man.

The axis of rotation through O can be at any place between the couple of forces, as can be seen in Fig. 2.8. Even in such cases, the resultant torque is the same as in (2.3):

$$M = Fd_{\perp 1} + Fd_{\perp 2} = F(d_{\perp 1} + d_{\perp 2}) = Fx.$$

It is important now to analyze the situation in which the forces that constitute the couple have lines of action not perpendicular to the bar, as illustrated in Fig. 2.9. Note that d is the distance between the points of application of the forces

Fig. 2.7 Two forces of equal magnitude but opposite direction are applied at the ends of a bar. The system rotates about the axis through O (perpendicular to this page) at the center of the bar. The binary arm is represented by x

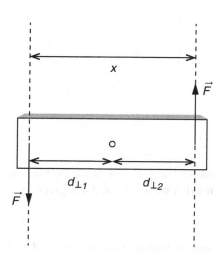

Fig. 2.8 A binary with axis of rotation through O not centered in relation to the forces **F** applied at the ends of the bar

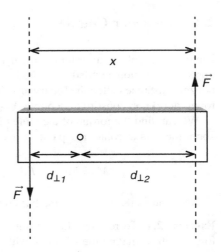

Fig. 2.9 Couple of forces **F** not perpendicular to the bar

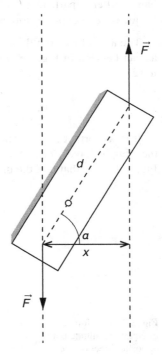

on the bar. Observe, however, that the arm of the binary is still the distance x. The relation between x and d is given by the following expression:

$$x = d \cos \alpha,$$

and the torque of this couple is $M = Fx$.

2.4 Torque Due to Two or More Nonparallel Forces

2.4.1 Resultant of Two Nonparallel Forces Applied on a Body and Its Line of Action

We have already seen that one of the ways to determine the net torque on a body is by summing the torques due to each force separately. By another method, we use the resultant force and its point of application, or better, its line of action and the correspondent arm.

In the study of torque, it is clear that the point of application of a force is of fundamental importance. Actually, if the line of action of the force is determined, the problem is solved, since the arm of the force corresponds to the perpendicular distance between this line and the axis of rotation.

Observe that the effect of a force on a body does not change if its point of application is changed, since it stays on the line of action of the force. Using this property, the resultant force can be determined, dislocating the vectors on their lines of action to work as shown in Fig. 2.10 for the case of two nonparallel coplanar (which are in the same plane) forces.

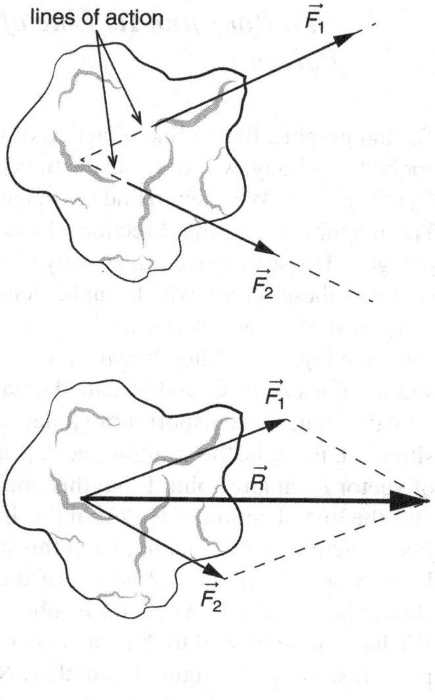

Fig. 2.10 Resultant force **R** is obtained by the rule of parallelogram, summing two forces whose points of application on the body do not coincide. Note that the rule is applied, first, by dislocating the points of application of forces to the same point. The line of action of the resultant is also determined by obtaining its torque about any axis

Fig. 2.11 Rule of polygon
applied to the forces F_1, F_2,
and F_3 to obtain the
resultant R. The polar rays
which connect the point P to
the head of each vector were
drawn in sequence

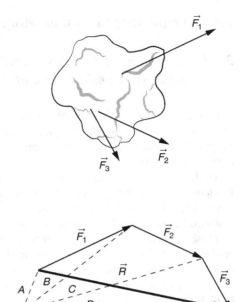

2.4.2 Resultant of Two or More Nonparallel Forces Applied on a Body and Its Line of Action: Method of Funicular Polygon

To find graphically the line of action of the resultant of two or more coplanar forces applied on a body, we construct the funicular polygon. Consider the forces F_1, F_2, and F_3 in Fig. 2.11. We want to find the magnitude of the resultant R and its line of action. The magnitude and the direction of the resultant R can be obtained by the rule of polygon. To use this method, we only need to draw straight lines parallel to the lines of action of these forces, which can be done with the help of two squares. The first step consists in the transportation of the vectors to add them by the rule of polygon, as shown in Fig. 2.11. After drawing the resultant, we adopt a point P and connect it with straight lines A, B, C, and D, called polar rays, to the vertices of the polygon.

After that, we transport these polar rays, using the same direction as the original situation. For this, the straight line A is drawn in a way to intercept the line of action of vector F_1 at any point. From this point of intersection, we draw the line B until it cuts the line of action of F_2. From this intersection, we draw line C until it meets the line of action of F_3. Finally, from this point, we draw line D. At the intersection of lines A and D, the line of action of the resultant R should pass and, furthermore, should be parallel to the vector R, obtained by the rule of parallelogram (Fig. 2.11). We have to be careful to draw always parallel lines, when transporting vectors, and polar rays from one figure to another. Note that sometimes the line of action of the resultant can pass outside the object subjected to the forces. The resultant of three forces applied on a body by the method of funicular polygon is shown in Fig. 2.12.

Fig. 2.12 Construction of funicular polygon to determine the line of action of the resultant of the three coplanar forces

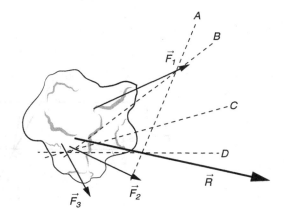

Once the magnitude and direction of the resultant force and its line of action have been determined, the torque due to this resultant can be calculated about any axis of rotation. This can be done after obtaining the arm $d_\perp$ and using (2.1) and specifying the direction of rotation.

Example 2.2 Consider a bar of 38 cm in length subjected to three coplanar forces with magnitude $F_1 = 5.1$ N, $F_2 = 12.6$ N, and $F_3 = 11.8$ N, as illustrated in the figure of Example 2.2. Determine the resultant force vector by the method of polygon and then apply the method of funicular polygon to obtain the line of action of the resultant force. Then, find the torque of the resultant force about the axis of rotation through O.

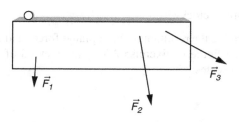

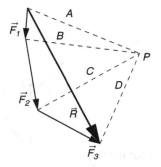

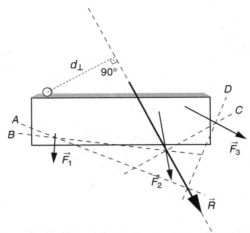

The first thing to do is to transport the force vectors to do a vector addition. We obtain for the magnitude of the resultant the value $R = 26.3$ N.

Next, from the point P, we draw the polar rays and then transfer them to the diagram of forces. The point of intersection between the first and the last polar rays defines the point where the line of action of the resultant force must pass. Then, the arm of the resultant can be obtained using the scale:

$$d_\perp = 16.5 \text{ cm} = 0.165 \text{ m}.$$

Hence, the net torque due to the resultant force about the axis at O is

$$M_R = -Rd_\perp = -(26.3 \text{ N})(0.165 \text{ m}) = -4.3 \text{ N m}.$$

Direction of rotation: clockwise

Exercise 2.5 Determine the resultant of the coplanar forces and its torque about the axis through O of the figure of Exercise 2.5 by the method of funicular polygon. Utilize the following scales:

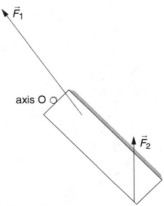

1 cm in the drawing = 10 N for the force.
1 cm in the drawing = 20 cm for the distance.

2.5 Rotational Equilibrium

For a body to be in rotational equilibrium, the sum of the torques produced by all the forces acting on the body must be zero. This is the condition used by the two-pan balance. The torque of the weight of an object in one of the pans of the balance makes it fall (rotate), if there is nothing in the other pan. As the calibrated mass is placed on the other pan, the torque due to its weight begins to equilibrate the balance. When the weights on both pans are the same, the magnitude of both torques will be the same and the net torque will be zero, due to opposite signs, and the balance will be in equilibrium.

Example 2.3 John and Mary are playing on a seesaw, as can be seen in the figure of Example 2.3. John, with a mass of 20 kg, is seated 2 m from the pivot point, called the fulcrum. At what distance from the fulcrum Mary, with a mass of 30 kg, has to sit so that the board will be in horizontal equilibrium?

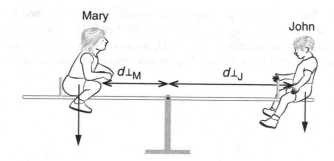

The sketch shows the situation:

To be in equilibrium, the sum of torques of weights of Mary and John must be zero, that is,

$$M_M + M_J = 0,$$

$$M_J = -(\text{Johns mass})gd_{\perp J} = -(20\,\text{kg})(10\,\text{m/s}^2)(2\,\text{m}) = -400\,\text{N m}.$$

Hence, in equilibrium, $M_M = +\,400$ N m $= (30\text{ kg})(10\text{ m/s}^2)d_{\perp M}$.
Then, $d_{\perp M} = (400$ N m$)/(300$ kg m/s$^2) = 1.33$ m.

As Mary weighs more than John, she has to sit closer to the fulcrum of the seesaw than John for the seesaw to be in horizontal equilibrium, that is, at 1.33 m from the pivot point O.

Exercise 2.6 You want to construct a mobile with four ornaments and three light rods with negligible mass as shown in the figure of Exercise 2.6. The distances in centimeters and the mass of one of the ornaments in grams are indicated in the figure. Determine the masses of ornaments A, B, and C, considering that the mobile should be in equilibrium.

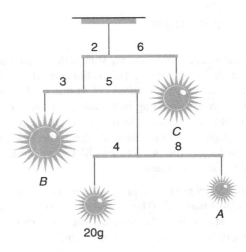

Example 2.4 The forearm of the figure of Example 2.4 is maintained at 90° to the upper arm. An object weighing 15 N is held in the hand. The distance between the object and the axis of rotation through O, located at the elbow joint between the ulna and the humerus, is 30 cm. The biceps muscle is attached at 3.5 cm from the joint O.

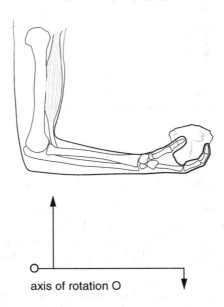

axis of rotation O

(a) Find the torque produced by the weight of the object.
(b) Find the upward force that the biceps muscle exerts on the forearm (made up of the ulna and the radius). Consider only these two forces and that the forearm is in equilibrium.

(a) $M_W = -(15 \text{ N})(0.30 \text{ m}) = -4.5 \text{ N m}$.

(b) If the forearm is in rotational equilibrium, the magnitude of the torque due to the muscle force must be equal to that calculated in (a) with the opposite sign, that is, $M_b = 4.5 \text{ N m} = F_b(0.035 \text{ m})$.

Hence, $F_b = (4.5 \text{ N m})/(0.035 \text{ m}) = 128.6 \text{ N}$. Note that the muscle force is 8.6 times larger than the weight of the object.

Exercise 2.7 In Example 2.4, the weight of the forearm and hand was neglected. Consider now that the weight of the forearm plus the hand of 20 N is applied at 15 cm from O. Find:

(a) The torque due to the weight of the forearm-hand
(b) The muscle force exerted by the biceps muscle

2.6 Answers to Exercises

Exercise 2.1 (a) $M = 0.60 \text{ N m}$; (b) $M = 2.80 \text{ N m}$; (c) $M = 5.30 \text{ N m}$; the same result is obtained when the angle between the outstretched arm and the body is $30°$, either upward or downward since $d_\perp$ is always equal to $d\sin30°$: (d) $M = 0.30 \text{ N m}$; (e) $M = 1.40 \text{ N m}$; (f) $M = 2.65 \text{ N m}$. The direction of rotation depends on the arm, either the left or the right arm.

Exercise 2.2 The progressive difficulty is due to the increase in the magnitude of torque which is a consequence of the increase in the distance of application of total weight (head-neck + torso + arms-forearms-hands and a part of the abdomen) to the axis of rotation at the lower abdominal region. As the arms-forearms-hands are brought closer to the head, the torque becomes larger. To overcome it, a larger torque with the opposite direction of rotation should be exerted, which demands larger muscle force.

Exercise 2.3 As the angle of the set thigh-leg-foot with the ground (horizontal) becomes smaller, the torque becomes larger, because the distance $d_\perp$ is larger, and the weight of the set is applied a little above the knee. As the leg is elevated, the torque will be less because the distance $d_\perp$ will be less. If the torque is larger, this must be equilibrated by another torque of opposite sign produced by muscle force. Therefore, the larger the torque, the larger the muscle force should be in order to maintain equilibrium. Bending the legs, the torque due to the weight will be much less, because the distance $d_\perp$ will shorten and the muscle needs to exert a much smaller force to reach the final posture of sarvanga-asana.

Exercise 2.4 $M = 20 \text{ N m}$.

Exercise 2.5 $m_A = 10 \text{ g}$; $m_B = 50 \text{ g}$; $m_C = 26.7 \text{ g}$.

Exercise 2.6 (a) $M = -3.0 \text{ N m}$; (b) the torque of the force exerted by the biceps muscle must equilibrate the sum of torques due to the weight of the object plus the weight of the forearm-hand. So, $F_b = 214.3 \text{ N}$.

Chapter 3
Center of Gravity

The weight of a body is an ever present force, and its consideration is of basic importance to any analysis of strength and of structural equilibrium. This chapter deals with the displacement and the rotation that this force can produce. To consider the effects it causes, we represent this force acting on a given point of the body, called the center of gravity of the body.

3.1 Objectives

- To discuss the concept of center of gravity of a body
- To determine the center of gravity of a body
- To discuss the stability of the human body
- To classify the state of equilibrium of a body as stable, unstable, or neutral

3.2 Weight and Center of Gravity

An extensive body can be imagined as composed of a very large number of very small pieces, as tiny as a cell, if this body is, for example, a piece of the human body. The resultant weight of this body will correspond to the sum of the gravitational forces that act on each of these small pieces. There is a point where we can consider that the resultant weight is applied, called the center of gravity, C.G. This fact can be interpreted as if all mass of the body is concentrated at this single point and, hence, this is the point of the application of the weight force. For homogeneous bodies of regular shape, the C.G. is in its geometric center. An object will be in equilibrium when suspended or supported by a force whose line of action passes through this point, the C.G. But one can ask: why it is so important to know the body's position of the center of gravity?

To respond to this question, let us begin by analyzing the situations illustrated in Fig. 3.1. Various bodies in this figure are in situations that we can ask what

E. Okuno and L. Fratin, *Biomechanics of the Human Body*, Undergraduate
Lecture Notes in Physics, DOI 10.1007/978-1-4614-8576-6_3,
© Springer Science+Business Media New York 2014

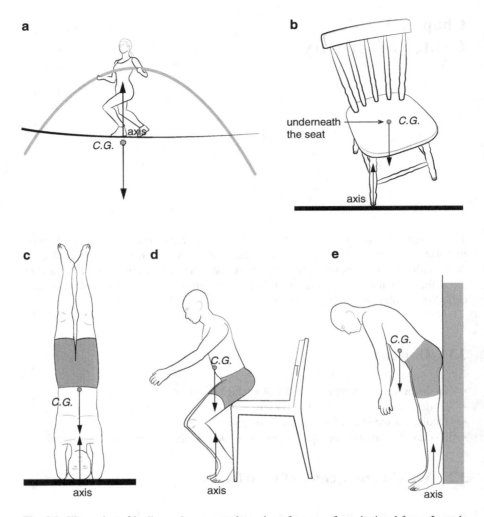

Fig. 3.1 Illustration of bodies under, or not, the action of torque of gravitational force. In each situation, the center of gravity C.G., the weight, the support point seen as an axis about which the body can rotate, and its respective normal force are represented. (**a**) Acrobat, (**b**) inclined chair, (**c**) yoga's headstand posture (sirsasana), (**d**) a person rising from a chair, (**e**) a person bending over to touch his toes with heels and back against a wall

will happen next: they will remain at the same position or will rotate or fall. The center of gravity of each body is represented in Fig. 3.1 as well as the point of support.

The conclusion for each situation is obtained by considering that the weight force of each body acts at its center of gravity and can exert a torque about the point of support, considered as an axis of rotation. As the weight force is always vertical

Fig. 3.2 (**a**) Illustration of
the projection (X) of the
center of gravity, C.G., of a
person standing erect, over
the area delimited by the
position of his or her feet.
(**b**) The same over an area of
support delimited by the tip
of the feet

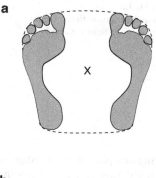

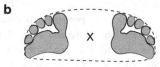

and downward (more precisely, directed toward the center of earth), we just need to
see if its lever arm is zero or not. If it is different from zero, there is torque, and
hence rotation which can be clockwise or counterclockwise. When the line of action
of the weight force passes through the axis of rotation, the lever arm is zero and
there will be no torque, and the body will remain at the same position in static
equilibrium. It is worthwhile to note that the support force or the normal force does
not exert torque, since it acts just at the point of support and therefore the lever arm
is null. However, the support guarantees the equilibrium against translational
motion because the net force applied to the body constituted by the weight and
normal force is zero.

For a body to be in rotational equilibrium, an imaginary vertical line, passing
through its C.G., must pass through the area delimited by the support points.
In the case of a person standing erect with both feet equally sustained by the
ground, the area that delimits the points of support involves both feet, as illustrated
in Fig. 3.2a. Figure 3.2b shows the area of support of the same person on the tip of
the feet. The stability will be greater the larger is this area. This is the reason that it
is difficult to remain in equilibrium standing on tiptoe or on the tiptoe of a single
foot, as the area of support is much smaller.

An important observation related to the position of the center of gravity is that it
can be located outside the body, since it is a function of its mass distribution, as will
be seen later on. This was illustrated in Fig. 3.1a, b for an acrobat and an inclined
chair, respectively.

Observe further that if the center of gravity is below the point of support, a
situation of great stability is achieved, as the torque caused by the weight produces
an oscillation that, when damped, leads to equilibrium with the point of support and
the center of gravity aligned vertically. Figure 3.3 shows a plastic bird with its
wings well opened, in equilibrium, supported by its beak. It is due to the lead

Fig. 3.3 Bird in
equilibrium, with the center
of gravity below its body

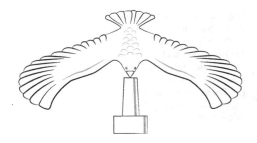

masses placed at the edge of the wings, which displaces its C.G. to outside of the
body, immediately below the beak, more or less at the height of the lead masses.

Now it is possible to understand why a baby does not begin by walking, but
crawls on all fours: one of the reasons is because, with the hands and feet on
the ground, the area where the vertical line passing through the C.G. may fall is the
greatest possible. Besides this, its C.G. will stay closer to the ground in the posture
of crawl rather than in standing, reinforcing the control of equilibrium.

Persons who practice yoga know that the lotus posture (padmasana) is ideal for
meditation, as the body contact area with the ground is large and its C.G. stays near
the ground, allowing great stability.

Exercise 3.1 Analyze each situation illustrated in Fig. 3.1 and determine whether
the body will remain in the represented position or will rotate. In the case of
rotation, specify whether the direction is clockwise or counterclockwise.

Exercise 3.2 Explain why it is impossible to bend over to touch the toes without
bending the knees if, initially, the heels and the back are against the wall. Discuss,
justifying, how this exercise can be done.

Example 3.1 Describe the movements of the center of gravity of a person's body,
initially seated, as this person rises from a chair.

When seated on a chair, the line of action of the weight force, which passes
through the center of gravity, is found to be near the geometric center of the area
delimited by the legs of the chair, which is the area of support, plus the area of the
feet on the ground. To rise, as the line that passes through the C.G. must intercept
the base support that now will be the feet, the person has to bend forward. This can
be facilitated, if the arms are outstretched forward.

3.3   Practical Method to Locate the Center of Gravity

After justifying the importance of knowing the position of the center of gravity, let
us learn how to locate it in any body. Initially, let us consider a practical method
also called the method of the plumb line.

Fig. 3.4 Representation of
a practical method to
determine the center of
gravity of a sheet of paper.
Its initial and final positions
are shown

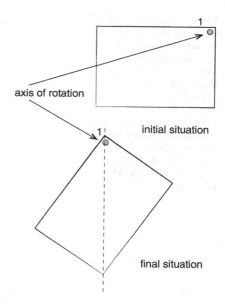

This method, illustrated in Fig. 3.4, will be applied to a flat body, such as a sheet of paper. Let us consider a rectangular and homogeneous sheet of paper. At the beginning, we choose an axis of rotation perpendicular to this sheet and fix it to a wall with a thumbtack, with the possibility to rotate. When the sheet is loosened, it will rotate and the final equilibrium situation is reached. Then, two questions arise:

(a) Why does sheet rotate when it is released?
(b) Why does sheet stop rotating and remain in a certain position at the final equilibrium situation?

To answer such questions, we have to consider the torque of the weight force about an axis of rotation defined at the point where it is fixed. Certainly, we can consider that the center of gravity of the sheet is not at the axis of rotation, but at some distance from it, as shown in Fig. 3.5. Being so, there will be the lever arm of the weight, whose torque will rotate the sheet. This is the answer to the first question.

To explain why the sheet stops rotating and reaches the final situation, we should remember that if there is no rotation, there is no torque, or better, the torque becomes zero. Knowing that the torque is given by the product of weight by the lever arm $d_\perp$, and the weight did not change, we conclude that $d_\perp$ became zero. This happens when the line of action of the weight force and therefore the C.G. of the body is on the vertical line which passes through the axis of rotation as can be seen in Fig. 3.5. This is the answer to the second question.

We have drawn a dotted vertical straight line at the final situation passing along the axis of rotation in Fig. 3.4. The C.G. must be found on this line. The precise

Fig. 3.5 Representation of the proposed experiment for the determination of the C.G., showing the applied forces and the effect of torque which produces counterclockwise rotation of the sheet from the initial to final situation. At the final situation, the torque is zero. In the repetition of the experiment, the sheet is suspended through axis 2 of rotation

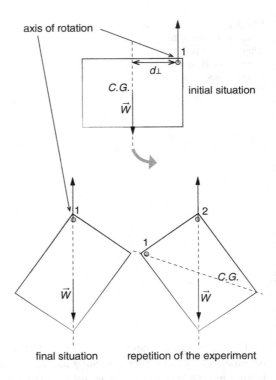

determination of the C.G. is achieved by repeating the previous procedure, fixing the sheet of paper through the second axis of rotation, and drawing another vertical line passing through the axis as shown in Fig. 3.5. The center of gravity must also lie along this second line. The only way for one point to fall on two different lines is to be located at their intersection. This point corresponds to the C.G. of the body.

In short, the method of the plumb line used to locate the center of gravity, C.G., of a body consists in suspending it from a point, and, at equilibrium, a vertical line which corresponds to the line of action of the weight force is drawn. This procedure is repeated one or more times, suspending the body from a different point each time. The C.G. must lie on both lines of action, and the only point which belongs to both lines is that of the intersection. For two-dimensional (three-dimensional) objects, it has to be suspended by at least two (three) different axes of rotation.

It is interesting to point out that, as the C.G. is the point of the body where its weight is considered to be applied, we shall be able to equilibrate the body by hanging or supporting it from a point such that the line of action of the force that supports it passes through its C.G.

The C.G. of the human body or parts of it has been obtained using cadavers by the practical method of the plumb line.

Exercise 3.3 Cut out any flat figure as illustrated in the figure of Exercise 3.3, from cardboard, and locate its center of gravity by the method of the plumb line.

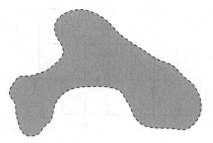

Exercise 3.4 Locate by the plumb line method the center of gravity of a hanger and explain why it is in equilibrium when hanging on a bar in the wardrobe.

Example 3.2 Threading a thread through a wedding ring as shown in the figure of Example 3.2, we see that it will be equally in equilibrium, independent of the place where the thread passes. Explain why.

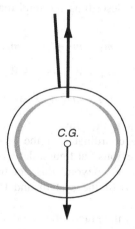

This occurs because the wedding ring has a symmetric shape and its C.G. is exactly at its center (does not belong to the ring). Wherever the thread passes through to hang it, the normal force there applied and the weight, at the C.G., will be always aligned.

3.4 Analytical Method to Locate the Center of Gravity

Figure 3.6 represents a flat body which can be imagined as being composed of a large number of small pieces, which will be designated elementary masses. The total mass, M^1, will correspond to the sum of the elementary masses m_1, m_2, etc.,

---

[1] Note that we are using the same letter M for the torque and total mass.

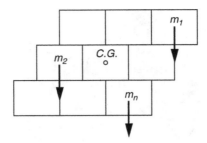

Fig. 3.6 Illustration of an extensive body composed of several elementary masses. The net weight of this body acts at the general center of gravity, C.G., represented in the figure. The *vertical downward arrows* represent the elementary weight forces. Only three of them have been drawn

and, hence, the total weight of this body, W, will correspond to the sum of weights of each of these pieces. The general center of gravity, C.G., is represented in the figure.

The mathematical relations describing the total mass and weight are given by (3.1), (3.2), and (3.3):

$$M = m_1 + m_2 + \cdots + m_n, \tag{3.1}$$

$$W = W_1 + W_2 + \cdots + W_n, \tag{3.2}$$

$$Mg = m_1 g + m_2 g + \cdots + m_n g, \tag{3.3}$$

where g is the acceleration due to the gravity.

Now let us determine the coordinates of the center of gravity, analytically. For this, let us analyze the previous flat figure, defining a point to be the origin of the system of coordinates (x, y). Beyond this, let us suppose that this body is suspended at the origin $(0, 0)$ of the system and that the y-axis is vertical as shown in Fig. 3.7.

As in the case of analysis by the practical method, the suspended body is subject to the torque, due to the weight, about the origin of system of coordinates which is also the axis of rotation in this case. The resultant or net torque can be calculated from the sum of each elementary torque, due to the action of weight of each elementary mass or from the torque of resultant total weight acting at the center of gravity where all the mass of the body is considered to be concentrated, whose coordinates we want to determine. In the case of Fig. 3.7, the lever arms of the elementary weights about the origin of the system of coordinates correspond to the x coordinates of each elementary mass.

The mathematical relation between the torque due to the total weight W of the body acting at the center of gravity and the sum of torques due to the elementary weights is given by (3.4):

$$W x_{C.G.} = W_1 x_1 + W_2 x_2 + \cdots + W_n x_n. \tag{3.4}$$

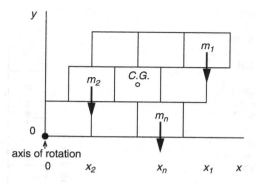

Fig. 3.7 Illustration of an extensive body composed of several elementary masses. In this case, the y-axis is vertical and parallel to the force (elementary weights) vectors and to the resultant weight acting at the center of gravity, C.G., whose coordinates are to be determined. The *vertical downward arrows* represent the elementary weights

Substituting the weights by the product of the masses and the acceleration of gravity, g, we obtain (3.5):

$$Mgx_{C.G.} = m_1 g x_1 + m_2 g x_2 + \cdots + m_n g x_n. \tag{3.5}$$

The coordinate $x_{C.G.}$ of the center of gravity is obtained, isolating it from (3.5). Observe further that the acceleration of gravity can be canceled on both sides of (3.5), if we consider it constant, at the location of the body. Then, we can write

$$x_{C.G.} = \frac{m_1 x_1 + m_2 x_2 + \cdots + m_n x_n}{M} = \frac{\sum_{i=1}^{n} m_i x_i}{M}. \tag{3.6}$$

The determination of the coordinate $y_{C.G.}$ of the center of gravity can be done similarly. For this, we have to rotate the system of coordinates without changing the disposition of the body about the x and y axes. Figure 3.8 illustrates the new situation where the x-axis is vertical, with the same direction of the line of action of the weight. Again, the resultant torque can be calculated from the sum of each elementary torque or from the torque of the total weight applied at the center of gravity. In the case of Fig. 3.8, the lever arms of elementary weights about the origin of the system of coordinates correspond to the coordinate y of each elementary mass.

Analogously to what has been done to determine the coordinate $x_{C.G.}$ of the center of gravity, we can write (3.7), (3.8), and (3.9):

$$Wy_{C.G.} = W_1 y_1 + W_2 y_2 + \cdots + W_n y_n, \tag{3.7}$$

$$Mgy_{C.G.} = m_1 g y_1 + m_2 g y_2 + \cdots + m_n g y_n, \tag{3.8}$$

Fig. 3.8 Illustration of the
same extensive body of
Fig. 3.6, but with a rotation
of x and y axis, in a way that
the x-axis is now vertical
and parallel to the
elementary weights and to
the resultant weight applied
at the center of gravity, C.
G., whose coordinates we
want to determine. The
downward arrows represent
some elementary weights

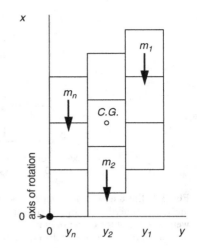

$$y_{C.G.} = \frac{m_1 y_1 + m_2 y_2 + \cdots + m_n y_n}{M}. \tag{3.9}$$

An analogous analysis performed to locate the coordinates $x_{C.G.}$ and $y_{C.G.}$ of the center of gravity could be done for a third coordinate, z, orthogonal to the coordinates x and y. This would complete the determination of the coordinates of the center of mass of an extensive body, with volume. Such a coordinate should be given by (3.10):

$$z_{C.G.} = \frac{m_1 z_1 + m_2 z_2 + \cdots + m_n z_n}{M}. \tag{3.10}$$

The expressions (3.6), (3.9), and (3.10) for the coordinates of the center of gravity correspond to the center of mass of the body. This identity is due to the fact that in the process of deduction of the coordinates of the center of gravity, the acceleration, g, due to the gravity was eliminated in the mathematical expressions. As already written, this can only be done if the value of g can be considered constant, which is true for small bodies compared to the size of earth. When this is not the case, the center of gravity does not correspond to the center of mass.

An important observation should be made about the origin of the system of coordinates. To obtain the coordinates of the center of mass, we have placed the origin at the stipulated axis of rotation, and we have based our analysis on the torque of the weight force. The mathematical equations obtained involve, however, only the elementary masses and their respective coordinates, limiting them to the spatial distribution of the mass of a body. In this case, it is possible to choose any origin for the system of coordinates, arbitrarily. Then, it is recommended, for convenience, that the choice should be done to simplify the calculations. One such example would be to choose an origin such that the significant part of the lever arms of the elementary weights should be zero for one of the coordinates. Observe, moreover, that the coordinates of the elements of mass can be positive or negative.

Example 3.3 Find the coordinates of the C.G. of the figure of Example 3.3:

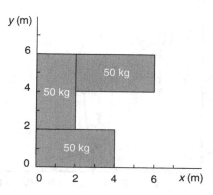

$$x_{C.G.} = \frac{(50\,\text{kg})(2\,\text{m}) + (50\,\text{kg})(1\,\text{m}) + (50\,\text{kg})(4\,\text{m})}{150\,\text{kg}} = 2.33\,\text{m}$$

$$y_{C.G.} = \frac{(50\,\text{kg})(1\,\text{m}) + (50\,\text{kg})(4\,\text{m}) + (50\,\text{kg})(5\,\text{m})}{150\,\text{kg}} = 3.33\,\text{m}.$$

In this case, the center of gravity does not belong to the body.

Exercise 3.5 Find the coordinates of the C.G. of the figure of Exercise 3.5:

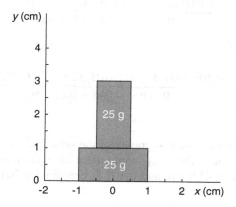

Example 3.4 Consider a seated person whose lower leg is vertical, as shown in the figure of Example 3.4. The coordinates of the center of mass of the thigh, leg, and foot are listed in the table of Example 3.4. The masses of thigh, leg, and foot are, respectively, 0.106 M, 0.046 M, and 0.017 M, with M the total mass of the body. Find the center of gravity of the set thigh/leg/foot of this person in this posture.

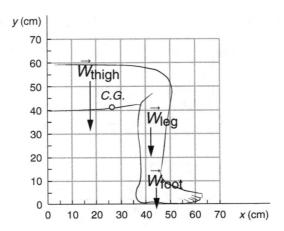

Table of Example 3.4

Parts of body	Coordinates of C.G.	
	x (cm)	y (cm)
Thigh	17.3	51.3
Leg	42.5	32.8
Foot	45.0	3.3

$$x_{C.G.} = \frac{(0.106M)17.3 + (0.046M)42.5 + (0.017M)45.0}{(0.106 + 0.046 + 0.017)M},$$

$x_{C.G.} = 26.9$ cm.

$$y_{C.G.} = \frac{(0.106M)51.3 + (0.046M)32.8 + (0.017M)3.3}{(0.106 + 0.046 + 0.017)M},$$

$y_{C.G.} = 41.4$ cm.

Exercise 3.6 Locate the center of gravity of the set thigh/leg/foot of Example 3.4 in the standing posture. The $x_{C.G.}$ coordinates of each part are, respectively, 42.5 cm, 42.5 cm, and 45.0 cm, and the $y_{C.G.}$ coordinates 76.6 cm, 32.8 cm, and 3.3 cm.

Exercise 3.7 Consider the following simulation of a human body with regular geometric shape and suppose that this is your body. Table 3.1 gives the approximate percentage mass of each part of the body. Represent the center of gravity of each part at its geometric center and find the coordinates of the simulated center of gravity using Fig. 3.9. Consider your height and total mass. To measure the coordinates (x, y, z) of the center of gravity about the origin of the system of

Table 3.1 Percentage
of total body mass

Parts of body	Mass (% of total body mass)
Head	6.9
Trunk-neck	46.1
Upper arms (2)	6.6
Lower arms (2)	4.2
Hands (2)	1.7
Thighs (2)	21.5
Legs (2)	9.6
Feet (2)	3.4
Total (whole body)	**100.0**

Fig. 3.9 Simulated body
of a man

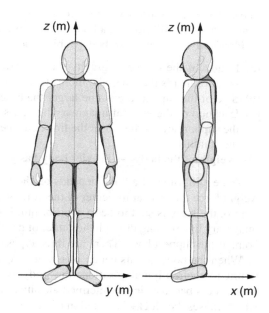

coordinates located on the ground between your feet, use a meter stick or a tape measure.

   The center of gravity of a person standing erect, with arms along the body, lies on a vertical line that touches the floor around 3.0 cm in front of the ankle joint. Its location is slightly in front of the second sacral vertebra or at 58 % of the height of a person above the ground. Since the location of the C.G. in human beings is relatively high above the area of support, their equilibrium is far from being stable. As the position of the C.G. depends on the distribution of mass, any modification of the shape of the body will change it. The center of gravity of men's body is slightly higher related to the support area than that of women, because men have wide shoulders and women large hips. During the last months of pregnancy, the C.G. of a woman's body is displaced forward, and to equilibrate, she walks by placing the torso and the head a little behind, seeming very proud of herself. People, who have had their leg or their arm amputated, have to retrain their equilibrium, because the position of their C.G. has been changed.

The difficulties in performing abdominals, with the arms much closer to the head and at the beginning of the yoga posture called sarvanga-asana with the legs closer to the ground, were discussed in Exercises 2.2 and 2.3. The main parameter involved in these situations is the change of the position of the C.G., which produces longer lever arms.

3.5   Stable, Unstable, and Neutral Equilibrium

The considerations already presented in this chapter allow us to classify briefly the conditions of equilibrium of a body related to the action of its weight force.

Physically, the degree of body stability depends on four factors, namely:

(a) Height of the center of gravity with regard to the ground—the nearer to the support area is the C.G., the more stable it is.
(b) Size of the support area—the larger the base area, the more stable it is.
(c) Location of the vertical imaginary line passing through the C.G. in relation to the support area—the closer the line to the center of the base area, the greater is the stability.
(d) Weight of the body—a heavier body has greater stability.

The equilibrium of a body depends on the existence, or lack, of a torque of the weight force that acts at its center of the gravity, as already discussed. If this torque is zero, the body is said to be in equilibrium. For this, it is only necessary that the imaginary line passing through the center of gravity passes also through the reaction point at the support base. There are three types of equilibrium.

When the body and its center of gravity are slightly displaced and both tend to return to their original position, the body is said to be in stable equilibrium. This occurs because, in the intermediate situation, the weight force exerts a torque which causes the clockwise rotation of the body, as can be seen in Fig. 3.10, which restores the original position. In this position, the vertical line passing through the center of gravity passes through the support base which, in this case, is not a point but the correspondent base area.

If the body and its center of gravity were displaced from their original positions and did not tend to return, but assume a new position, the body is said to be in unstable equilibrium. In this case, the torque of the weight force at the intermediate situation causes clockwise rotation, which will make the body assume a new position, which is the final one, as shown in Fig. 3.11. In the intermediate situation, the vertical line passing through the center of gravity falls outside the base support, rotating the body and driving it to the final position, caused by the torque of the weight force.

Neutral equilibrium refers to the displacement of the body and its center of gravity in a way that the torque due to the weight force is always zero about the support point. This means that the body will remain in any similar position in which it was placed. This situation is illustrated in Fig. 3.12.

Fig. 3.10 Body in stable equilibrium, since it will return to its original position when disturbed slightly, as indicated

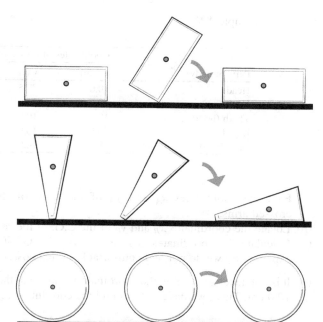

Fig. 3.11 Body in unstable equilibrium, as it will assume a new position when released after undergoing a small displacement as indicated

Fig. 3.12 Body in neutral equilibrium, as it will remain in a similar position when released, after a small displacement

Example 3.5 Consider an adult with 70 kg mass and 1.70 m height, seated on the floor, with crossed arms and outstretched legs, as shown in the figure of Example 3.5. It is not very easy to remain in this posture during a long time. The figure shows the lateral cut, and the table of Example 3.5 gives the *x* and *y* coordinates of the center of gravity of some segments, as well as the respective masses. The C.G. of head and of neck-trunk-crossed arms is vertically aligned at a distance of 81 cm from the sole of the feet.

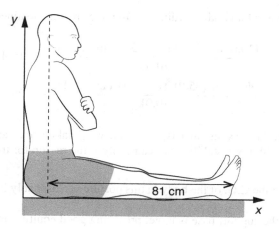

Table of Example 3.5

Parts of body	Coordinates of C.G.		Mass (kg)
	x (cm)	y (cm)	
Head	10.0	77.5	4.8
Neck-trunk-crossed arms	10.0	36.0	41.1
Thigh (both)	19.0	10.0	15.0
Leg (both)	60.0	10.0	6.7
Feet (both)	88.0	10.0	2.4

(a) Find the coordinates x_A and y_A of the C.G. of the set head-neck-trunk-crossed arms.
(b) Locate the coordinates x_B and y_B of the C.G. of the set thigh-legs-feet.
(c) Calculate the coordinates $x_{C.G.}$ and $y_{C.G.}$ of C.G. of this person's body and explain why we do not feel comfortable in this posture.

(a) It is not necessary to find the coordinate x_A, since all the segments are aligned at 10.0 cm. Then, we have only to find the coordinate y_A:

$$y_A = \frac{(77.5\,\text{cm})(4.8\,\text{kg}) + (36.0\,\text{cm})(41.1\,\text{kg})}{45.9\,\text{kg}} = 40.3\,\text{cm},$$

$x_A = 10.0$ cm.
(b) In this case, we do not have to calculate the coordinate y_B which is equal to 10.0 cm for all segments. Let us find, then, the coordinate x_B:

$$x_B = \frac{(19.0\,\text{cm})(15.0\,\text{kg}) + (60.0\,\text{cm})(6.7\,\text{kg}) + (88.0\,\text{cm})(2.4\,\text{kg})}{24.1\,\text{kg}} = 37.3\,\text{cm},$$

$y_B = 10.0$ cm.
(c) Now we have to find both coordinates, from the answers obtained in (a) and (b):

$$x_{C.G.} = \frac{(10.0\,\text{cm})(45.9\,\text{kg}) + (37.3\,\text{cm})(24.1\,\text{kg})}{70.0\,\text{kg}} = 19.4\,\text{cm},$$

$$y_{C.G.} = \frac{(40.3\,\text{cm})(45.9\,\text{kg}) + (10.0\,\text{cm})(24.1\,\text{kg})}{70.0\,\text{kg}} = 29.9\,\text{cm}.$$

Based on the last results, that is, $x_{C.G.}$ and $y_{C.G.}$, taking into account the four factors related with body stability, we can analyze the reason for the discomfort in this posture:

1. The height of the C.G. of the body in this posture is reasonably low, giving good equilibrium.
2. The size of the support base is large, providing good equilibrium.

3. However, the vertical line which passes through the C.G. is displaced from the center of the support base, being closer to the hips. Therefore, a general tendency in this posture is to bend forward, thus increasing the $x_{C.G.}$, in an attempt to make the vertical line through the C.G. approach the center of the base support.

Exercise 3.8 Now, analyze and justify the degree of difficulty in remaining lying on the floor with the legs lifted until the toes are pointed straight up, as illustrated in the figure of Exercise 3.8. This posture is the same as in Example 3.5, but with a 90° rotation. Which posture is easier, that of Example 3.5 or of this exercise?

3.6 Motion of the Center of Gravity

Consider a racket in translational motion, as shown in Fig. 3.13a. When a body experiences translational motion, every point of the body has the same displacement in any small time interval as its center of gravity. Therefore, the motion of the center of gravity of the body is called translational motion of a body.

However, when the same body rotates and whirls about an axis while in motion as in the case of a racket thrown by one person to another, as shown in Fig. 3.13b, only the center of gravity moves through a simple parabolic trajectory. Each elementary mass performs a different motion, and the trajectory is not as simple as that of the center of gravity.

When a diver jumps off a trampoline, he changes the shape of the body during the fall and each body part performs a different trajectory, complicated, but the trajectory of the center of gravity is described by a parabola, which can be written mathematically.

Fig. 3.13 (a) Translational motion of a racket. (**b**) Motion of a racket thrown by one person to another

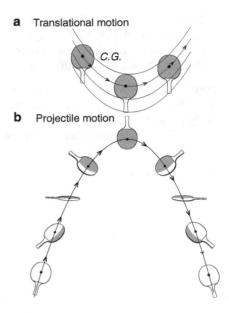

a Translational motion

C.G.

b Projectile motion

3.7 Answers to Exercises

Exercise 3.1 Situation (a) the acrobat does not rotate.

Situation (b) the chair rotates clockwise.

Situation (c) the yoga practitioner does not rotate and remains in the posture.

Situation (d) while the person keeps the vertical line passing through his C.G. on his support feet, he will not rotate.

Situation (e) when the vertical line passing through his C.G. extends beyond his feet, he will rotate counterclockwise.

When the C.G. is not along the same vertical line passing through the support point, there is a lever arm of the weight force and, hence, the torque will produce rotation. In cases in which the C.G. lies in the vertical line passing through the support point, the torque of the weight force is zero, since its lever arm is zero.

Exercise 3.2 This is the situation (e) illustrated in Fig. 3.1. At the instant in which the vertical line passing through his C.G. goes outside the support area, corresponding to the area of the feet in this case, the lever arm of weight force will no longer be zero and its weight will exert torque causing rotation, what could mean his fall. To bend over without falling, the person must maintain the line passing through his C.G. on the feet, moving the legs and hips backward.

Exercise 3.4 Depending on the material and shape, the C.G. of a hanger can be outside the hanger (in the air) or on it, but always below the point where it is supported when hanging in the wardrobe. This is a situation of great stability, since the torque of the weight force (weight + hanged clothes) will be always zero, when changed. Even displaced, the hanger oscillates and returns always to the equilibrium position.

Exercise 3.5 $x_{C.G.} = 0$ cm; $y_{C.G.} = 1.25$ cm.

Exercise 3.6 $x_{C.G.} = 42.75$ cm; $y_{C.G.} = 57.30$ cm.

Exercise 3.7 The location of center of gravity varies from person to person and depends on the height and weight of body parts.

Exercise 3.8 Analogously to what was said in Example 3.5, the vertical line passing through the C.G. of the set thighs-legs falls near the hips of the support areas. There is a tendency to bend the set thighs-legs to the side of the head, trying to make the vertical line fall closer to the center of the support area. This posture is relatively easier than that of Example 3.5, since in this case the vertical line passing through the C.G. of the body is closer to the center of the base area.

Chapter 4
Rotations

Rotational motions have their origin in net external torque or are modified by it. The difficulty in changing the state of rotation of a body depends not only on its mass but also on how this mass is distributed about the axis of rotation. For two bodies of the same mass, the body with the mass distributed closer to the axis of rotation will present less resistance to change in its rotational motion.

4.1 Objectives

- To investigate the physical quantities of moment of inertia and angular momentum
- To analyze the physical quantities that are conserved in rotations
- To relate the change in the angular velocity to modifications in the mass distribution of a rotational system
- To investigate these quantities in the human body

4.2 Moment of Inertia

In activities of athletes of different modalities, of circus artists, of ballet dancers, or of skaters, rotational motions are very common. Besides, all wheels of cars, of bicycles, and of many electric devices, such as mixers and ventilators, also work based on rotational motions. In this chapter, let us apply the concepts of physics to explain some of the characteristics related to rotational motions.

In translational motions, the state of motion of a body only changes if a nonzero resultant force F^1 acts on it. In other words, we can say that if a body is at rest,

---

[1] In this chapter we are using only the magnitude of vectorial quantities such as force, velocity, acceleration, torque, and angular momentum. Therefore, they appear written in italics, but not in boldface.

E. Okuno and L. Fratin, *Biomechanics of the Human Body*, Undergraduate
Lecture Notes in Physics, DOI 10.1007/978-1-4614-8576-6_4,
© Springer Science+Business Media New York 2014

it will continue at rest, and if it is moving at constant velocity, it will continue moving at constant velocity only if a resultant force F that we will call here simply force F (Newton's first law, already presented in Chap. 1) is not applied on it. On the contrary, if a force F is applied on the body, it will be accelerated with acceleration $a = \Delta v / \Delta t$, and the larger the magnitude of force F, the larger will be the acceleration acquired by the body (Newton's second law, already presented in Chap. 1). If we double the force applied on this body, the acceleration will double and, with triple the force, the acceleration is also tripled and so on. We can write mathematically (4.1) that F/a is constant for a given body. This constant is characteristic for a given body and is its mass m:

$$\frac{F}{a} = m. \tag{4.1}$$

This equation indicates that for a given applied force F, the larger the body mass, the less will be the acceleration acquired by it. Therefore, the mass of the body is a measure of body inertia, with obvious relevance for the difficulty in accelerating as well as decelerating an object. The jabiru is a bird found in the Pantanal of Mato Grosso, Brazil, which has a large body mass and to take off needs to run a lot to acquire enough velocity, and lands awkwardly, unlike other flying birds with less mass that move gracefully. Body inertia is directly proportional to its mass. Moreover, for translational motions, how the body mass is distributed is not important.

In rotational motions of a body about an axis, its state of motion is only modified if a resultant torque M acts on it. The difficulty here in changing its state of rotation depends not only on body mass but on how it is distributed about an axis of rotation. Equation (4.2) is equivalent to (4.1) for rotational motions:

$$\frac{M}{\alpha} = I, \tag{4.2}$$

with M the magnitude of net torque, I the moment of inertia, and α the angular acceleration.

The physical quantities of (4.1) and (4.2) that correspond are $M \leftrightarrow F$, $I \leftrightarrow m$, and $\alpha \leftrightarrow a$. As the mass m is related to the body's inertia in translational motions, the moment of inertia I is related to rotational inertia, which implies the difficulty in increasing or decreasing the angular acceleration. The moment of inertia increases with the increase of both mass and distance that characterizes the distribution of this mass about an axis of rotation, and hence, it will be larger as this distance increases. Therefore, the shape as well as the mass of a body determines how difficult it is to set it into rotation, as can be seen in Fig. 4.1.

All three types of wheels in Fig. 4.1 have the same mass, but their shapes are different, i.e., the mass composing them is differently arranged around the axis of rotation. The mass that composes wheel (a) is farther from the rotation axis than in wheel (b), and the wheel (b) mass further than wheel (c) mass. After a certain time

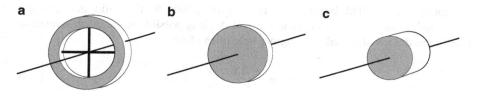

Fig. 4.1 Three wheels with the same mass, but with different shapes. The different mass distribution about an axis of rotation determines the specific difficulties to reach a given angular velocity, as well as to stop, if this is the case

Fig. 4.2 A particle of mass *m* moving in a circular path. It sweeps out an angle $\Delta\theta$ in a time interval Δt

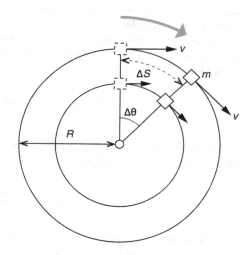

interval in applying the same torque, wheel (c) will have performed more revolutions than wheel (b) and the number of revolutions of wheel (a) will be the smallest, as it has the greatest moment of inertia *I*.

Let us see how we can write a mathematical expression that allows us to define moment of inertia. For this, let us analyze a particle of mass *m* rotating with an angular velocity ω, about an axis of rotation, shown in Fig. 4.2 (larger circumference). The angular velocity, described by (4.3), is given by the rate at which θ changes with time *t*:

$$\omega = \frac{\Delta\theta}{\Delta t}. \tag{4.3}$$

The angular velocity ω in the SI must be expressed in radians per second (rad/s). Note that 2π rad $= 6.28$ rad $= 360°$. We remember that $\omega = 2\pi f$, with *f* the rotational frequency, measured in rotations per second (rps), which receives a special name hertz, shortened Hz, in the SI and hence 1 Hz $= 1/s = s^{-1}$.

The angular displacement $\Delta\theta$ corresponds to the displacement ΔS of the particle along the circular path divided by the radius *R*. Considering that the translational velocity, also called linear velocity v, corresponds to the ratio between the traveled distance ΔS and the time interval Δt, it is possible to establish a relation between the translational and angular velocities (4.4):

$$\omega = \frac{\Delta S}{R \Delta t} = \frac{v}{R}. \tag{4.4}$$

If another particle of mass m, rotating about the same axis, but in a smaller circumference, sweeps out the same displacement $\Delta \theta$ in the interval Δt, it means that this particle has the same angular velocity, but its linear velocity is smaller. Equation (4.4) shows that to maintain ω constant, if v is smaller, it is because R is proportionally smaller.

The energy associated with a particle in motion is called the kinetic energy, K. It depends on the mass m and on the linear velocity v of the particle, calculated from (4.5):

$$K = \frac{mv^2}{2}. \tag{4.5}$$

Substituting the velocity v by the product ωR obtained from (4.4), we can write for the particle's rotational kinetic energy, K_{ROT} (4.6):

$$K_{ROT} = \frac{mR^2 \omega^2}{2}. \tag{4.6}$$

Comparing (4.5) and (4.6), it is possible to establish an analogy between rotational and translational motion. We note that, as we have substituted the translational velocity v by the rotational angular velocity ω, the translational mass m was substituted by the product mR^2 for rotation. This product is called the particle's moment of inertia, and its unit in the SI is kg m^2.

The moment of inertia I of one single particle of mass m rotating about its axis at distance R is, then, given by (4.7):

$$I = mR^2. \tag{4.7}$$

The moment of inertia of an extensive solid body about an axis of rotation is the sum of moments of inertia of each elementary mass or particle of mass m_i, which composes the body about its axis of rotation:

$$I = m_1 R_1^2 + m_2 R_2^2 + m_3 R_3^2 + \cdots + m_n R_n^2 = \sum_i m_i R_i{}^2. \tag{4.8}$$

Observe that the mass distribution is more significant than the total mass because R appears squared in (4.7) and (4.8).

4.3   Moment of Inertia of Regularly Shaped Uniform Solids

Many times, a body of complex shape can be simulated by the composition of uniform and geometric regular-shaped bodies. We have already adopted this to evaluate the center of gravity of a human body (Fig. 3.9). Therefore, it is very useful to know the moments of inertia of some of these bodies. The moments of inertia about a predefined axis of rotation for some solids of total mass M can be calculated with the mathematical technique of integrals. As such calculations are outside the scope of this book, here we will present the results of such calculations and discuss them. In the figures, I_y means moment of inertia about the y-axis and so on. Observe that we are using the same symbol M for the torque and for the total mass of a solid body.

4.3.1   Radius of Gyration

For bodies of any shape, we can always find a point where the total mass M of the body as a whole is considered to be concentrated, without changing its moment of inertia about a given axis. This point is at a distance k from the axis of rotation and is called the radius of gyration of the body about this axis. Note here that, in general, the mass of a body cannot be considered as being concentrated at the center of gravity for the purpose of calculating the moment of inertia. The introduction of the radius of gyration is simply to facilitate the visualization since the values of k are given in tables for solids of uniform density and commonly encountered shapes.

(a) Figure 4.3 shows a disc of radius R and mass M that rotates about the y-axis passing through its center of gravity. In this case,

$$I_y = M\frac{R^2}{4} = Mk^2$$

from which we find that $k = R/2 = 0.5\,R$. For a disc with radius $R = 0.20$ m and $M = 1$ kg, we obtain $I_y = 0.01$ kg m^2. Note that if the figure is rotated 90°, the disc will be horizontal and the axis of rotation now will be also horizontal and called x, but the situation is equivalent and, hence, the formula for the moment of inertia is the same.

(b) Figure 4.4 shows a disc of radius R and mass M that rotates about the x-axis passing through the center of gravity. In this case,

$$I_x = M\frac{R^2}{2} = Mk^2$$

from which we find that $k = \frac{R}{\sqrt{2}} \approx 0.71R$. Observe that the body is the same as in the previous example; only the axis of rotation has been changed, and the moment

Fig. 4.3 (**a**) Disc of radius R and mass M rotating about the y-axis. (**b**) Representation of the same situation with the radius of gyration. Here, all the mass M of disc (*gray small sphere*) is concentrated at the distance k of the axis of rotation

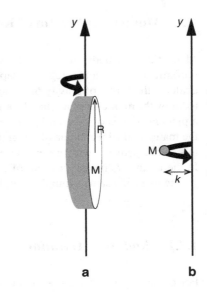

Fig. 4.4 (**a**) Disc of radius R and mass M rotating about the x-axis. (**b**) Representation of the same situation with the radius of gyration

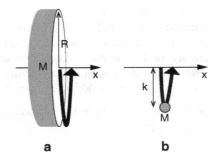

of inertia that depends on the mass distribution about an axis has increased. For a disc of radius $R = 0.20$ m and $M = 1$ kg, we obtain $I_x = 0.02$ kg m^2, i.e., double that of I_y.

(c) Figure 4.5 shows a solid sphere of radius R and mass M that rotates about any axis passing through its center of gravity. The moment of inertia in this case will be the same, independent of the axis of rotation, due to the spherical symmetry:

$$I = M\frac{2R^2}{5} = Mk^2$$

in this case $k = \sqrt{\frac{2}{5}}R \approx 0.63R$. A sphere can represent a crouched person, clasping the knees during a revolution after a jump from a springboard. For a sphere with $R = 0.20$ m and $M = 1$ kg, we obtain $I = 0.016$ kg m^2.

Fig. 4.5 (**a**) Solid sphere of
radius R and mass M that
rotates about an axis
passing through its center.
(**b**) Representation of
the same situation with
the radius of gyration

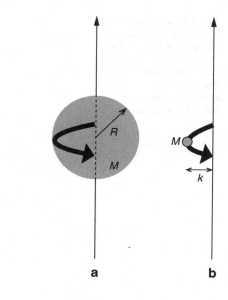

a   b

Fig. 4.6 (**a**) Solid cylinder
with radius R and mass
M that rotates about the
x-axis. (**b**) Representation
of the same situation with
the radius of gyration

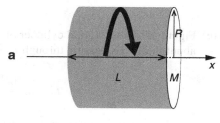

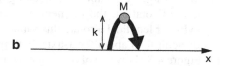

(d) Figure 4.6 shows a solid cylinder of radius R, length L, and mass M that
rotates about the x-axis passing through its center of gravity. The moment of
inertia is

$$I_x = M\frac{R^2}{2} = Mk^2$$

in this case, $k = 0.71R$. An upper arm or a forearm can be represented by a
cylinder. Note that in this case the moment of inertia does not depend on the
length L. For a cylinder of radius $R = 0.05$ m, $L = 0.20$ m, and $M = 1$ kg,
we obtain $I_x = 0.0013$ kg m^2.

Fig. 4.7 (a) Solid cylinder
with radius R, length L,
and mass M that rotates
about the y-axis passing
through the center of
gravity. (b) Representation
of the same situation with
the radius of gyration

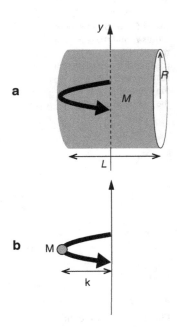

(e) Figure 4.7 shows a solid cylinder of radius R, length L, and mass M that rotates
about the y-axis passing through its center of gravity. In this case,

$$I_y = M \frac{3R^2 + L^2}{12} = Mk^2$$

and $k = 0.50R + 0.29L$. For a cylinder with $R = 0.05$ m, $L = 0.20$ m, and
$M = 1$ kg, we obtain $I_y = 0.004$ kg m^2. Note that changing the position of the
axis of rotation, the moment of inertia changed and now it depends on the
cylinder length. To rotate this same cylinder about the y-axis is 3.0 times more
difficult than about the x-axis, both passing through the center of gravity.

(f) Figure 4.8 shows a solid cylinder of radius R, length L, and mass M that rotates
about an axis passing through one of the bases. It is possible to obtain the
moment of inertia of a body, when the axis of rotation is parallel to an axis
passing through its center of gravity by applying the theorem of parallel axis.

4.3.2   Parallel Axis Theorem

The moment of inertia of a body can be calculated in relation to any axis of
rotation previously defined. The moments of inertia in relation to the axis of
rotation passing through its center of gravity have great applications and are
calculated and tabulated. Observe that the moments of inertia of bodies of

Fig. 4.8 Solid cylinder of radius R, length L, and mass M that rotates about the y-axis passing through one of the bases

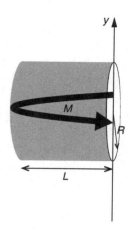

Figs. 4.3, 4.4, 4.5, 4.6, and 4.7 are all about an axis passing through its center of gravity. There is a very useful and simple relation between the body moment of inertia about an axis passing through its center of gravity $I_{\text{C.G.}}$ and the moment of inertia I of this body about another axis parallel to the first. With M the total body mass and $d = L/2$, the distance between both parallel axes, the relation is given by

$$I = I_{\text{C.G.}} + Md^2. \tag{4.9}$$

Using (4.9) we can deduce several expressions of interest as, for example, to obtain I_y of Fig. 4.8 from I_y in Fig. 4.7 which will be $I_{\text{C.G.}}$. In the case of Fig. 4.8,

$$I_y = M\frac{3R^2 + L^2}{12} + M\frac{L^2}{4} = Mk^2.$$

For a cylinder with radius $R = 0.05$ m, $L = 0.20$ m, and $M = 1$ kg, we obtain $I_y = 0.014$ kg m^2, hence, 3.5 times more difficult to rotate about this axis than about another y-axis passing through its center of gravity.

Example 4.1 Is it possible to explain, without calculating, why tightrope walkers on ropes or steel cables at height use long poles?

The explanation comes from the fact that, in holding a long pole horizontally, the moment of inertia of the body plus the pole about the y-axis passing through the center of gravity increases, and consequently the rotational inertia increases, i.e., it becomes more difficult to rotate the body; in other words we have increased the equilibrium of the body. Of course, concentration is essential. In our daily life, we usually open our arms outstretched to better equilibrate when we walk on a narrow wall.

Exercise 4.1 Determine the moment of inertia of a forearm plus a hand with a total mass of 3.0 kg, about a rotation axis according to the figure of Exercise 4.1. Suppose that the forearm plus the hand has the form of a cylinder with length equal to 0.38 m and radius of 0.03 m. Find the value of radius of gyration and the moment of inertia for each case.

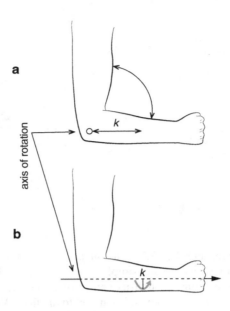

4.4 Moment of Inertia of the Human Body

The distribution of bones, fat, and muscles in the human body varies from subject to subject, depending also on age. All of this, allied to the fact that parts of the body or the whole body does not have regular geometric shapes, raises difficulties in determining the moment of inertia by mathematical calculation. For each segment and for the whole body of a standard man, the moment of inertia about a given rotation axis was obtained using cadavers, mathematical model simulations, and via photographic methods. This approach is very much exploited by sportsmen in activities like running, changing the angle of the leg in relation to the thigh or in ornamental jumping, by doubling up and clasping the knees to modify the moment of inertia in order to obtain the desired performance.

To define the moment of inertia, it is necessary to specify first the axis of rotation. The human body rotates, when it is free of support, about three axes, called the principal axes. These are mutually perpendicular lines that pass through the center of gravity corresponding to the posture of the body. These axes are called transverse, anteroposterior, and longitudinal. Figure 4.9 shows these axes. The moment of inertia related to each of these axes is called the principal moment of inertia. Depending on the shape of the body, the moment of inertia acquires a different value. The estimated values of the moment of inertia for a man of 70 kg mass and 1.70 m height are as follows:

$I_{\text{longitudinal with arms along the body}} = 1.0\text{--}1.2 \text{ kg m}^2$.
$I_{\text{longitudinal with horizontally outstretched arms}} = 2.0\text{--}2.5 \text{ kg m}^2$.

Fig. 4.9 Principal axes of rotation for a human body that rotates freely, without support

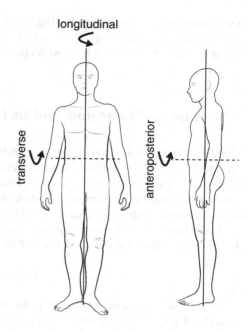

$I_{\text{transverse of man standing with arms along the body}} = 10.5–13.0 \text{ kg m}^2$.
$I_{\text{transverse of man crouched with the hands on the knees}} = 4.0–5.0 \text{ kg m}^2$.
$I_{\text{anteroposterior of man standing with arms along the body}} = 12.0–15.0 \text{ kg m}^2$.

Example 4.2 The moment of inertia of a person with 70 kg mass and 1.70 m height about the longitudinal principal axis with the person's arms along the body is $I = 1.0$ kg m^2 as illustrated in Fig. 4.9. (a) Determine the radius of gyration about the longitudinal principal axis. (b) Suppose that an equivalent geometric model to this person is a cylinder of 70 kg mass and 1.70 m height. Find the radius of this cylinder.

Observe that the radius of gyration is the distance from the place where the total mass of the body can be concentrated to produce the same moment of inertia for the given axis of rotation.

(a) $1.0 \text{ kg m}^2 = (70 \text{ kg})k^2$; isolating k, we obtain

$$k = \sqrt{\frac{1.0\,\text{kg m}^2}{70\,\text{kg}}} = 0.12\,\text{m}.$$

(b) In this case $I = MR^2/2$; hence, $R = \sqrt{\frac{2I}{M}} = 0.17\,\text{m} = 17\,\text{cm}$. This model is reasonable.

Exercise 4.2 A person with 70 kg has a radius of gyration equal to 0.17 m about the main longitudinal axis when she or he is with outstretched arms at an angle of 90° in relation to this axis as illustrated in Fig. 4.9. Determine the moment of inertia of this person about the main longitudinal axis. Compare the degree of difficulty to rotate about this axis when she or he is with opened or closed arms.

4.5 Angular Momentum and Its Conservation

Consider an ice skater rotating with a certain angular velocity about her or his longitudinal axis with horizontally outstretched arms. What happens if she or he closes their arms, hugging the axis of rotation? Why does an athlete change the shape of his body to perform an ornamental jump? To understand such questions, let us introduce a physical quantity called angular momentum L.

Every body rotating about an axis has an angular momentum L (4.10) given by the product of the moment of inertia I of the body and the angular velocity, ω:

$$L = I\omega. \tag{4.10}$$

Hence, the angular momentum depends not only on the mass but on its distribution about a rotational axis and the angular velocity. Its unit in the SI is kg m^2/s, with the radian (rad) of angular velocity dropped.

The angular momentum is conserved (the value does not change) as the net total torque due to the external actions on the body is zero. Then, if no external torques act on a rotating body, it will maintain its rotation indefinitely, conserving its angular momentum. If no external torques act on the body that is not in rotation, on the other hand, the body will not rotate and its angular momentum will continue zero.

Equation (4.10) shows that, as the value of L must be maintained constant, if I increases, ω must diminish and vice versa.

Example 4.3 Consider the wheel of a bicycle with radius R of 30 cm and mass M of 2.0 kg that is supposed to be concentrated at its rim. The linear velocity of the wheel is 5.0 m/s. For objects with this geometric shape, the moment of inertia about the axis passing through its C.G. can be calculated by $I = MR^2$. Find for the wheel (a) its moment of inertia, (b) the radius of gyration, (c) the angular velocity, and (d) the angular momentum.

(a) $I = (2.0 \text{ kg})(0.30 \text{ m})^2 = 0.18 \text{ kg m}^2$.
(b) $k = R = 30$ cm.
(c) $\omega = v/R = (5.0 \text{ m/s})/0.30 \text{ m} = 16.7$ rad/s.
(d) $L = I\omega = (0.18 \text{ kg m}^2)(16.7 \text{ s}^{-1}) = 3.0 \text{ kg m}^2/\text{s}$.

Exercise 4.3 A car with 1,500 kg mass is moving with a velocity of 144 km/h = 40 m/s in a circular racecourse with 50 m radius. Obtain the angular velocity and the angular momentum about the center of the racecourse. In this case use the equation $I = mR^2$, since the auto can be approximated as a particle with mass m, rotating about the axis of rotation as shown in Fig. 4.2, as the radius is very large.

Fig. 4.10 Representation of a ballerina who is able to increase her rotational velocity by a maneuver of closing her arms, as the rotational angular momentum has to be conserved

4.5.1 Angular Impulse

Now we introduce another physical quantity: the angular impulse (4.11). It is given by the product of external net torque and the duration of the action and is responsible for the variation of angular momentum, $\Delta L = L_{\text{final}} - L_{\text{initial}}$, of a body:

$$\text{Angular impulse} = \text{Torque}_{\text{ext}}\Delta t = \Delta L. \qquad (4.11)$$

The rotational motions of the human body occur about the axis that passes through its center of gravity. In this case, an important observation is that the weight force acting on the body's center of gravity does not produce torque and, hence, does not change its angular momentum. Torques are originated by impulse forces that will introduce or change the angular momentum of body. In case that the impulse force is not applied, the body will maintain its state of rotation, that is, its angular momentum is conserved.

Figure 4.10 shows a ballerina in evolution. She is able to increase her rotational velocity simply by closing her arms. When she is with outstretched arms, she has a greater difficulty to rotate, i.e., she has more rotational inertia (moment of inertia) than with closed arms. Therefore, since the angular momentum is conserved, when she closes her arms, an increase in the angular velocity must occur, according to (4.10).

Example 4.4 Consider an ice skater with 60 kg mass and radius of gyration of 0.15 or 0.11 m about the main longitudinal axis when she is with opened or closed arms, respectively. Considering that her angular velocity is 6 rad/s with opened arms, determine her angular velocity when she closes her arms. Calculate the number of turns, supposing that she remains with this angular velocity (closed arms) during 30 s.

Opened arms: $L = I\omega = mk^2\omega = (60 \text{ kg})(0.15 \text{ m})^2(6 \text{ s}^{-1}) = 8.1 \text{ kg m}^2/\text{s}$.

Closed arms: the angular momentum is conserved after her maneuver. Hence, $8.1 \text{ kg m}^2/\text{s} = (60 \text{ kg})(0.11 \text{ m})^2\omega$:

$$\omega = \frac{8.1\,\text{kg}\,\text{m}^2/\text{s}}{(60\,\text{kg})(0.11\,\text{m})^2} = 11.15\,\text{rad/s}.$$

If in 1 s she turns through an angle of 11.15 rad, in 30 s she will turn 334.5 rad. As 1 turn $= 2\pi$ rad $= 6.28$ rad, the number of turns she will complete is 53.

Exercise 4.4 A diver with 65 kg is standing, facing the swimming pool. He puts his arms behind him and at the moment of the jump receives an angular impulse of reaction on his feet exerted by the springboard. The rotation of his arms also introduces angular momentum in the process of jumping. His rotational motion will occur about the main transverse axis. Knowing that when he leaves the springboard, his radius of gyration is 0.5 m and his angular velocity is 3.5 rad/s, determine the diver's angular velocity at the moment when he passes from the outstretched posture to the curved, clasping his knees, decreasing his radius of gyration to 0.28 m. Find the diver's angular momentum during the jump. The greater the angular momentum reached, the larger is the number of turns in the air that he can perform before touching the water's surface.

Another example of angular momentum conservation occurs during the spike in a volleyball game. During a jump, the rotation of arms with large angular velocity in the direction of the ball is necessarily accompanied by the leg's rotational motion (in opposite direction). The legs, being heavier than the arms and, therefore, with smaller angular velocity, their angular momentum compensates the arm's angular momentum. Figure 4.11 illustrates this situation. The initial angular momentum is zero and, hence, during the spike, must remain zero, which explains the reason for compensation.

Fig. 4.11 Example of angular momentum conservation in a spike in volleyball game. The angular momentum introduced by the rotation of the arms is compensated by the rotational motion of legs

It is worthwhile to make a final observation related to the conservation of total angular momentum in the absence of external torques. There are motions of legs, arms, and hips that allow transference of rotation from a main axis to another axis. This is observed during aerial evolutions in circus acrobatics, in ornamental jumps, and in several Olympic gymnastics modalities where, during a jump with apparent rotation about a main transverse axis, the athlete also presents a screw rotation. In the case of total angular momentum equal to zero, its conservation requires possible rotations which occur at least in two body segments, where one segment balances the angular momentum of the other segment.

4.6 Variation of Angular Momentum

When an external nonzero resultant force acts on a body, producing an external torque, the variation of the angular momentum occurs during the time interval corresponding to the duration of action. The force of reaction from a surface is very common in producing the angular impulse with which the rotation begins in aerial trials. This situation is shown in Fig. 4.12.

Example 4.5 Consider that the diver of Fig. 4.12 has a mass of 60 kg and a radius of gyration $k = 0.45$ m about the main transverse axis. The force of reaction from the springboard on the diver has a magnitude of 600 N and its lever arm $d = 0.15$ m. The force of reaction acts 0.5 s during the jump. Determine the angular momentum introduced on the diver as well as the angular velocity with which the diver leaves the springboard.

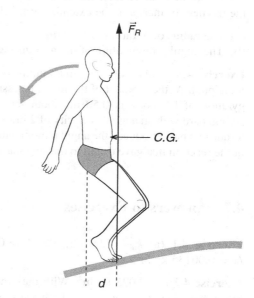

Fig. 4.12 Representation of a diving jump in which the force of reaction from the springboard on the diver is responsible for the introduction of the angular momentum which will be conserved during the jump

Torque of the reaction force: $M_{FR} = (600 \text{ N})(0.15 \text{ m}) = 90 \text{ N m}$.

Angular impulse due to this torque: $I_A = (90 \text{ N m})(0.5 \text{ s}) = 45 \text{ N m s}$.

This impulse corresponds to the angular momentum introduced: $L = 45 \text{ kg m}^2/\text{s}$.

$L = I\omega = mk^2\omega$, that is, $45 \text{ kg m}^2/\text{s} = (60 \text{ kg})(0.45 \text{ m})^2\omega$.

$$\omega = \frac{45 \text{ kg m}^2/\text{s}}{(60 \text{ kg})(0.45 \text{ m})^2} = 3.7 \text{ rad/s}, \quad \omega = 3.7 \text{ rad/s}.$$

Exercise 4.5 Consider the rotation of a body with 8 kg mass with the radius of gyration $k = 0.2$ m and angular velocity $\omega = 3$ rad/s. Find the moment of inertia and the angular momentum. Repeat the exercise in case each of the quantities is doubled, changing each time only one of the variables.

Exercise 4.6 In the table of Exercise 4.6 are given the moments of inertia about the main longitudinal axis for a dancer in two different moments of evolution and respective angular velocities. Obtain the angular momentum in each situation and try to justify the maneuver performed by the dancer.

Table of Exercise 4.6

Evolution	$I_{C.G.}$ (kg m^2)	ω (rad/s)
A	3.0	20.0
B	5.0	12.0

Exercise 4.7 The arm, with 3.5 kg mass, of a volleyball player moves with an angular velocity of 18 rad/s about his shoulder, during the spike. Considering that the moment of inertia of the extended arm is 0.40 kg m^2, determine the following:

(a) The radius of gyration of the arm
(b) The angular momentum of the arm during the spike

Exercise 4.8 A diver of 70 kg mass jumps from the springboard, facing the swimming pool. At the moment of the jump, he assumes such a posture that the radius of gyration of his body is 0.50 m, about the main transverse axis. He leaves the springboard with an angular velocity of 2 rad/s. Determine the magnitude of external torque that has introduced the angular momentum on the diver, knowing that the force of the reaction that gave origin to this angular momentum had a duration of 0.4 s.

4.7 Answers to Exercises

Exercise 4.1 (a) $k_a = 22.0$ cm and $I_a = 0.145$ kg m^2; (b) $k_b = 2.1$ cm and $I_b = 0.00135$ kg m^2.

Exercise 4.2 $I = 2.023$ kg m^2. With the arms opened, the degree of difficulty is twice that when a person is with both arms near the body and along the body.

Exercise 4.3 $\omega = 0.8$ rad/s; $L = 3 \times 10^6$ kg m^2/s.

Exercise 4.4 $\omega = 12.1$ rad/s; $L = 61.5$ kg m^2/s.

Exercise 4.5 For $m = 8$ kg: $I = 0.32$ kg m^2; $L = 0.96$ kg m^2/s.
For $m = 16$ kg: $I = 0.64$ kg m^2; $L = 1.92$ kg m^2/s.
For $k = 0.4$ m: $I = 1.28$ kg m^2; $L = 3.84$ kg m^2/s.
For $\omega = 6$ rad/s: $I = 0.32$ kg m^2; $L = 1.92$ kg m^2/s.

Exercise 4.6 Situation A: $L = 60$ kg m^2/s.
Situation B: $L = 60$ kg m^2/s.
The dancer was with closed arms in A and opened in B.

Exercise 4.7 (a) $k = 0.34$ m; (b) $L = 7.2$ kg m^2/s.

Exercise 4.8 $M_{FR} = 87.5$ N m.

Chapter 5
Simple Machines

Detailed analysis of simple machines, such as levers, pulleys, and inclined planes, allows us to appreciate and to respect human ingenuity. As machines, they are projected to do work and to facilitate human action.

5.1 Objectives

- To apply the principle of the lever and the equilibrium conditions for a rigid body
- To classify the levers
- To investigate several systems of force in the human locomotor equipment, identifying the existing types of levers
- To analyze systems with pulleys and inclined planes which, together with levers, constitute what are called simple machines

5.2 Simple Machines

We consider a machine as being a mechanism projected to perform a particular task, facilitating or enabling human action. Among the machines considered simple are levers, pulleys, and inclined planes. A mechanical analysis of levers allows us to understand and to determine the muscle efforts which are exerted on the bones to support or to move the resistance originated by the weights of such structures, whether or not the additional external loads are summed. The pulleys are part of physiotherapy equipment with properties that allow a change in the direction of forces and to multiply the effect of actions, balancing larger forces with smaller ones. Possibly, the inclined plane is the oldest simple machine.

E. Okuno and L. Fratin, *Biomechanics of the Human Body*, Undergraduate
Lecture Notes in Physics, DOI 10.1007/978-1-4614-8576-6_5,
© Springer Science+Business Media New York 2014

5.3   Work Done by a Force

In physics, the meaning of the work is well defined and it differs from the meaning in daily use. When an object, subjected to the action of a force, moves from one point to another, we say that this force performed work. Always, work is done by a force. Work τ, done by a force F exerted on the object, is defined as the product of the force F by the displacement d of this object in the direction of the applied force (5.1). Hence, the simple fact that a force is applied on an object does not imply that work is performed, since it is necessary that there is a displacement in the direction of this force. In the case of a person standing motionless and carrying a huge weight on his or her back during a long time, neither this weight force, nor this person does any work, according to physics.

$$\tau = Fd. \tag{5.1}$$

The basic unit of work in the SI is newton·metro (N m). Observe that this unit is the same as that of torque but with a different meaning. In the case of work, there is a special unit in the SI that is joule, abbreviated 1 J = 1 N m.

The concept of work is very important, since it is associated with the concept of energy. A body has a quantity of energy correspondent to its capacity to do work, as defined above. Hence, the unit of energy is also joule.

When a force of 1 N = 1 $kg\ m/s^2$ is applied on a body and it moves 1 m, we say that this force has performed work of 1 J.

5.4   Levers

A lever can be represented by a long stick or by a bar, which, by the action of forces, can or cannot rotate about an axis or a pivot, called the fulcrum. In biomechanics, the concept of the lever appears in each set composed of articulation (axis of rotation), bones (bar), and muscles (muscle force).

To represent a lever we distinguish three forces: the action force, F_A, the resistance (load) force, F_R, (which opposes our action), and the reaction force at the fulcrum. The action or applied force is the force exerted by the muscle and the resistance force is the load. The action and the resistance forces give rise to torques about the axis. The perpendicular distance between the axis and the line of action of the resistance force is called the resistance arm, d_R, and the equivalent distance to the line of action of the action force is called the action arm, d_A.

There is a classification of levers that allows us to identify them in a great number of situations, showing that they are more present in our daily life than we imagine. The levers are classified in three categories.

Fig. 5.1 Representation
of the first class lever,
denoting the position of
the axis of rotation,
perpendicular to this page,
the force of action and of
resistance, and their arms.
An example of a first class
lever used to lift a heavy
object is also represented

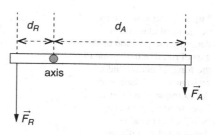

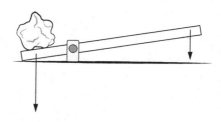

Fig. 5.2 Representation
of the second class lever,
indicating the position of
the axis of rotation in
relation to the force of
action and of resistance,
and the arms of these forces.
The sketch of a second class
lever utilized to transport
heavy loads is also
exemplified

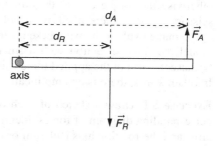

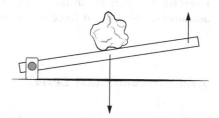

5.4.1 First Class Levers

All of the systems in which the fulcrum or pivot (axis of rotation) is between
the points of application of the action force and that of the resistance belong to this
category. Figure 5.1 shows a generic representation of a first class lever. The force
of reaction at the fulcrum is not drawn in this figure as in Figs. 5.2 and 5.3. The mass

Fig. 5.3 Representation
of the third class lever,
denoting the position of the
axis of rotation in relation to
the force of action and of
resistance, and their arms.
An example of a third class
lever used to support a
heavy object is also
represented in a diagram

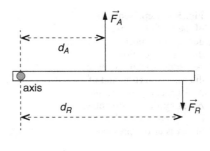

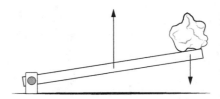

of the bar was not taken into account, being considered negligible, because
otherwise the torque due to the weight of the bar must be considered a pair of
scissors, a pair of pliers, a seesaw, a nail puller etc. are examples of first class levers.

To make explicit the way these levers are constructed, we represent in a typical
model the force of action with a smaller magnitude than that of the resistance
because the arm of the applied force was designed to be larger than that of the load.
In other words, there is an amplification of the effect of the action force.

Exercise 5.1 Draw a sketch of each one of the quoted first class levers and the
corresponding diagram of forces, identifying not only the forces involved but their
arms and the rotation axis (fulcrum or pivot).

Exercise 5.2 Indicate in the example to the right of Fig. 5.1 the corresponding
distances of the arms of force of action and of resistance.

5.4.2   Second Class Levers

In this category of levers, the load is between the fulcrum and the action force.
Its generic representation can be seen in Fig. 5.2.

Normally we want to amplify the effect of our action force by the use of this type
of lever. Therefore, in a typical representation, the magnitude of the action force
is drawn smaller than that of the load, because the arm of the applied force has been
designed to be larger than the arm of the load. Second class levers are probably the
more numerous. Some examples are: screwdrivers, doorknobs, tire-irons, door
keys, wheelbarrows, can opener, pedals, flywheels etc.

Exercise 5.3 Draw a sketch of each of the quoted second class levers and the corresponding diagram of forces, identifying the involved forces, their arms and the rotation axis (fulcrum or pivot).

5.4.3 Third Class Levers

In these levers, the action force is applied between the fulcrum and the load. Figure 5.3 shows the typical representation of levers of this category.

Observe that, in the representation of third class levers, the applied force has a larger magnitude than that of the load because the arm of the action force is smaller than that of the load. In other words, the effect of the action force is reduced. Tweezers and salad or pasta tongs are good examples of this type of lever. Third class levers are predominant in the human body. These lever systems are designed for increasing the speed of motion rather than increasing load capabilities.

Exercise 5.4 Draw a sketch of each of the quoted third class levers and the corresponding diagram of forces, identifying the involved forces, their arms, and the rotation axis (fulcrum or pivot).

5.4.4 Mechanical Advantage

In representations utilized for levers, we tried to make explicit the configurations that amplify or reduce the effect of the action force. We note that in the case of third class levers, the magnitude of the action force is greater than that of the resistance force. The contrary occurs in the case of second class levers. In case of first class levers, the magnitude of forces will depend on its construction or the way it is used. It is verified that, for the forces considered, the determinant parameter will be the arm of each force.

Let us imagine the situation of equilibrium of a lever, that is, in which there is equality between the torques of the action force and that of the resistance force (5.2):

$$F_A d_A = F_R d_R. \tag{5.2}$$

This equality shows that a force with a smaller magnitude can produce the same torque as that of a larger magnitude force, just by increasing its arm. Then, we define what is called mechanical advantage, MA, which corresponds to the ratio between the resistance force and the action force (5.3), that is, how many times the resistance force is larger than the action force:

$$MA = \frac{F_R}{F_A}. \tag{5.3}$$

So, as the mechanical advantage becomes larger, the necessary effort for a determined task becomes smaller and vice-versa. The mechanical advantage of levers can also be written as the ratio between the arm of the action and the arm of the load (5.4), which is obtained using (5.2) and (5.3):

$$MA = \frac{d_A}{d_R}.$$ (5.4)

Equation (5.4) allows us to design or to analyze more easily a lever, as we do not need to know the magnitude of the forces, but just their arms. In designing a tool, it is sufficient to analyze adequately the position of the hands. In the human body, for example, if the position of the ligaments with respect to the bones and the position of the resistance force are identified, we will easily identify the respective arms of the forces about a given articulation. The type of lever more commonly found in the human body is of the third class. In it, the resistance arm is always longer than the arm of the muscle force and the mechanical advantage can be 0.1 or even smaller.

Exercise 5.5 Suppose that you use a bar with a length of 75 cm, with the axis at one end, to lift a load of 8 kg placed at the opposite end. Find the force of action you must exert at a distance of 15 cm from the axis. Classify the type of lever.

Exercise 5.6 A nut is placed at 2 cm from the hinge of a nutcracker. Supposing that you exert a force of 25 N at a point located at 15 cm from the hinge, find the resistance force exerted by the nut. Classify the type of lever.

Exercise 5.7 Suppose that you exert an action force of 300 N at a point 50 cm distant from the base point (axis) of a lever. Find the load you are able to lift if it is placed at 10 cm from the axis, on the opposite side. Classify the type of lever.

5.5   Levers in the Human Body

In the movements performed by human bodies, mechanical levers are present. The bones (segments) work as a rigid structure (bar) on which forces are exerted. The articulations correspond to the fulcrum (pivot) and the muscles and ligaments provide the forces to move the loads. In biomechanics, forces are represented by vectors whose lines of action depart from the point of muscle attachment into the bone. The vectors follow the muscle direction, but not necessarily the anatomical direction of the entire muscle. Figure 5.4 shows the biceps muscle and the force it exerts. In Example 2.4, this force was considered perpendicular to the radius (horizontal), a bone of the forearm.

5.5.1   The Locomotion Equipment

Human locomotion equipment is constituted of around two hundred bones, articulations and muscles. Besides structural function allowing support, and to contain and to give form to soft tissues, such as muscle mass, fat, and skin,

Fig. 5.4 The force exerted
by the biceps muscle is
represented by the vector
that departs from the point
of muscle attachment to the
bone and follows the muscle
direction

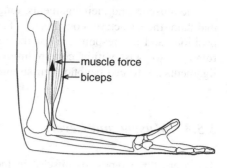

Fig. 5.5 Models of
articulation (**a**) ball
and socket type;
(**b**) hinge type

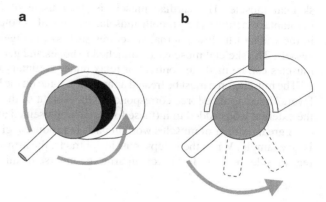

it enables the human being to move himself and to displace with considerable
effort, that is, performing mechanical work.

The shape of bones and their internal structure combine lightness and strength.
Articulations, well lubricated, slide smoothly without friction, ensuring the most
diverse movements.

5.5.2   Articulations and Joints

Joints are places where the bones touch. Some are immovable, connecting the bones
firmly (joints of the skull). Articulations are movable joints that enable movements.
Two examples of articulations are shown in Fig. 5.5:

- Ball and socket type, such as the shoulder joint that allows rotational motion of
  the arms
- Hinge type, such as the knee joint and the elbow joint that allows motion in a
  plane

The articulated bones have soft cartilage at their articulation extremity. In
addition, the region is filled with viscous fluid that guarantees good lubrication.

The muscles and their ligaments, tough fibrous strands, together with the bones maintain the structure constituted of the bones and the articulations at their positions and are responsible for the movement of this structure, allowing it to rotate or to twist within certain limits. Considerable damage occurs when the ligaments are forced beyond their strength limit and rupture.

5.5.3   Muscle and Levers

There are three types of muscles in the body: the cardiac, the smooth, and the skeletal muscle. The cardiac muscle is the muscle of the heart which has an involuntary action. The smooth muscle, usually of involuntary action, is located in the walls of hollow internal structures such as the digestive tract and the blood vessels. The skeletal muscles are attached to bones and generate movements. These muscles can be made to contract or relax under voluntary control.

The muscle force must be treated as an action force on the bar (bone segment) of the lever. The resistance force corresponds to the weight of the segment plus the weight the external loads added to it (these forces are represented at their center of gravity).

A great number of muscles work in pairs to produce a given motion. Arm flexion is an example. When the biceps muscle contracts, the arm flexes and the triceps, the rear muscle, relaxes. To stretch an arm, the inverse of this occurs.

5.5.4   Identification of Levers in the Human Body

5.5.4.1   First Class Levers

In the body, this system of levers is often used to maintain posture or balance. One example of a first class lever in the body is found with the skull pivoting on the first vertebra, the atlanto-occipital articulation of the spine which acts as fulcrum. The weight of the head is the resistance force which is balanced by the force of extensor muscles (action force). The same principle is found in the intervertebral articulations for both seated and standing postures in which the weight of the trunk is equilibrated by the forces of the spinal erector muscle, acting on the vertebral axis.

5.5.4.2   Second Class Levers

This system provides mechanical advantage and is rarely found in the human body. The pushup is an example of a second class lever. In it, the fulcrum is at the toe-tip, the weight, at the body's center of gravity, corresponds to the resistance force and the action force is exerted by the arms during the pushup.

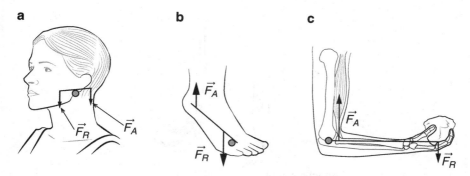

Fig. 5.6 Examples of levers in the human body. Levers of first, second and third class are represented in (**a**), (**b**), and (**c**), respectively. In all types, F_R is the weight and F_A is the muscle force

5.5.4.3   Third Class Levers

This class of lever is very common in the human body. In this case, the arm of resistance is always longer than the action arm, causing a mechanical disadvantage (mechanical advantage equal to 0.1 or smaller). On the other hand, a small shortening of a muscle causes a great arc of motion, being able to transport a relatively small load over a large distance. There are several examples: the deltoid acting on the glenohumeral articulation, the superficial flexor of the toes in the interphalangic articulations, the radial extensor of the wrist, the anterior tibia at the ankle articulation, and the brachial biceps at the elbow.

The first step in the analysis of a lever is to identify its axis of rotation. Then we should identify the lines of action of the muscle force (action) and of the resistance force (load). Finally, the arms of these forces are determined. Figure 5.6 shows three examples of levers in the human body, with one of each type. The forces of resistance (load) and of action (muscle) and the fulcrum are represented for each type of lever.

Exercise 5.8 Flex and stretch your arm, observing carefully the action of the brachial biceps and the triceps muscles. Represent each situation by the corresponding diagrams and classify the levers.

Exercise 5.9 Consider an arm under the action of the brachial biceps acting as a flexor in several postures, as illustrated in the figure of Exercise 5.9. Represent this muscle action in each case, identify its line of action, and determine the respective arm of the action force. Determine the angle in the case of maximum torque.

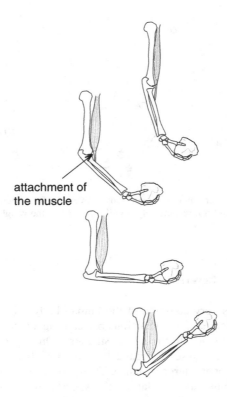

attachment of
the muscle

Exercise 5.10 The segment forearm-hand weighing 15 N with an additional weight of 50 N on the hand, at 25 cm from the elbow, is maintained at 45° with the humerus oriented vertically, as shown in the figure of Exercise 5.10. The center of gravity of the forearm-hand is located at 15 cm from the articular center of the elbow, and the flexor muscle is inserted at 3 cm from the articular center. Determine the intensity of the flexor muscle force to maintain this position and classify the type of lever. (sin 45° = 0.707)

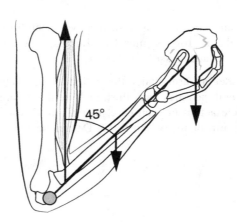

Exercise 5.11 A hand located at 30 cm from the elbow exerts a force of 50 N on the scale, as shown in the figure of Exercise 5.11. Consider that the triceps is attached inserted in the ulna at an angle of 90°, at 2.5 cm distance from the elbow. The weight of the forearm plus hand of 15.6 N is applied at its center of gravity at 15 cm from the elbow. Note that the force that the scale acts on the hand is of contact (normal) and its value is 50 N. The reaction force at the axis of rotation is not drawn. Determine the intensity of force in the triceps.

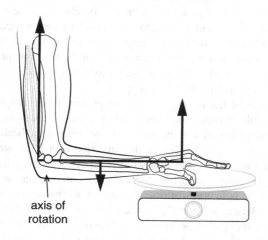

axis of
rotation

Exercise 5.12 A therapist applies on the forearm a lateral force of 80 N at 25 cm from the elbow, as illustrated in the figure of Exercise 5.12. The biceps is inserted in the radius at an angle of 90° and at a distance of 3 cm from the elbow. Determine the magnitude of the force on the biceps and the intensity of the reaction force exerted by the humerus on the ulna.

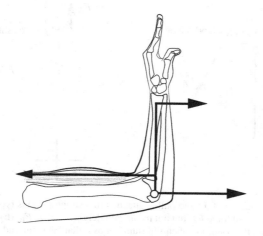

5.6   Pulleys

Up to here we have discussed the action of muscles on bones to perform any movement or even to sustain a certain structure in its place, when it is submitted to external loads. However, we should not forget situations in which external forces, such as traction that gives rise to torques and lever effects, are exerted by equipment in hospitals or even in fitness gyms. Usually these forces have a magnitude determined by the system of pulleys, which will be discussed next.

A pulley is a disc or a wheel with a groove around it where a rope, a cord, a cable or a chain passes whose function is to change the direction of the force. The disc has a central axis about which it rotates. The pulleys can be fixed or movable. The axis of a fixed pulley is attached to some support, while in a movable pulley one of the extremities of the rope is fastened to a support.

Several structures of the human body have properties of simple fixed pulleys. Tendons have the role of ropes and bone prominences, of discs (fixed pulleys). The lubricant fluids reduce the friction between the tendon and the bone to almost zero.

A fixed pulley allows you to change the direction of a force without, however, changing its magnitude. At one extremity of the rope, a resistance force F_R (load W) is applied and at the other, the action force F_A, as can be seen in Fig. 5.7a. If the object with weight W is in equilibrium, the sum of all the forces applied to it must be equal to zero. Hence, a force F_A of the same intensity as the weight W, with opposite direction, must also be applied to the object. As the rope changes the direction of the force, the force F_A to the right that will be applied by a person has the same magnitude as the force F_A to the left. Let us now demonstrate that $F_A = F_R$.

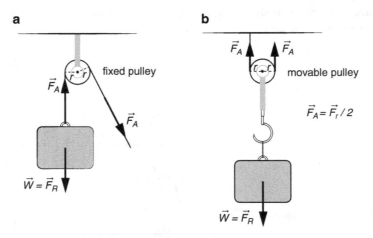

Fig. 5.7 (**a**) Representation of a fixed pulley attached to a structure. This type of pulley changes the direction of the applied force F_A, but not its magnitude, $F_A = W = F_R$. (**b**) Representation of a movable pulley. Here, the condition of equilibrium imposes that the magnitude of force F_A be half that of $W = F_R$, that is, $F_A = W/2$

Fig. 5.8 Representation of a combination of a fixed pulley with a movable pulley used to lift a load with a force of action $F_A = W/2$

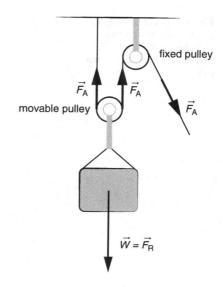

A fixed pulley behaves as a first class lever with equal arms r which is the radius of the wheel. If the system is in equilibrium, the torques caused by the forces must be equal:

$$F_A r = F_R r.$$

From this equation, we can conclude that $F_A = F_R$. In this case, the mechanical advantage is 1 because the magnitude of the action force is the same as the magnitude of the resistance force.

Figure 5.7b shows a movable pulley. It behaves as a second class lever, in which the arm of the action force is $2r$ and that of the resistance, r, taking as the axis the point where the force F_A is applied on the rope at the left. Therefore, if the system is in equilibrium we can write:

$$F_A 2r = F_R r.$$

From this equation, we find that $F_A = F_R/2$. Thus, the mechanical advantage in a movable pulley is 2. As can be seen, movable pulleys allow us to exert a force of magnitude smaller than would be required without it. It is worthwhile to note that, in our analysis, the weight of the pulley was not considered.

5.6.1 Combination of Pulleys

It is possible to project systems of complex pulleys that involve many pulleys and that certainly lead to a significant mechanical advantage.

Figure 5.8 shows a combination of a fixed with a movable pulley. As can be seen, in this case, the force of action is reduced to half of the resistance which is the

Fig. 5.9 Representation of
a pulley-block with two
movable pulleys and a
mechanical advantage equal
to four

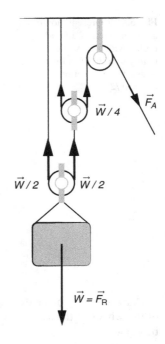

weight of the object that we want to lift, and the direction of the force has been changed.

The pulley-block, shown in Figs. 5.8 and 5.9, is the system of pulleys, constituted by independent movable pulleys that are supported by an even number of forces. In this case, as already demonstrated, each movable pulley works in a way that the action force necessary to maintain the system in equilibrium is half of that needed without the combination of pulleys.

In the pulley-block with N movable pulleys, the mechanical advantage is then given by:

$$\text{MA} = \frac{F_\text{R}}{F_\text{A}} = 2^N, \tag{5.5}$$

or more explicitly, we can say that the magnitude of the action force, F_A, is the magnitude of the resistance force, F_R, divided by 2^N. In the case of Fig. 5.8, the mechanical advantage is 2^1, that is, 2. On the other hand, in the case of Fig. 5.9, the mechanical advantage is 2^2, that is, 4, since $F_\text{A} = W/4$.

The system shown in Fig. 5.10 is made up by three pulleys: two fixed and one movable. Applying the condition of static equilibrium on the movable pulley, we verify that the vector sum of three upward forces must be the same as the weight of the suspended object. The weight of the wheel and of the ropes has been neglected. In the case in which we can consider the upward forces as vertical, we can say that the magnitude of the action force F_A in the figure is 1/3 of the weight of the hanging

Fig. 5.10 Combination of a
movable pulley (*below*)
with two fixed pulleys
(*above*) guaranteeing a
mechanical advantage
up to three

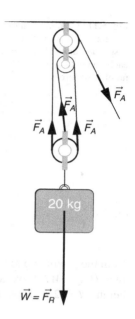

object, that is, $W = 3F_A$. The magnitude of force F_A is equal to 66.7 N and the mechanical advantage of the system is three.

In Fig. 5.10 the movable pulley is suspended in equilibrium by the action of three forces. On the other hand, Fig. 5.11 illustrates a combination of pulleys in which a configuration with two fixed pulleys and a support of the load with two movable pulleys is adopted. In this case, the same cable passes through all of the fixed and movable pulleys, giving rise to the four forces that support the load. In static equilibrium, these forces are of the same magnitude whose sum has the same value as the weight of the suspended load and the mechanical advantage is four. Therefore, the magnitude of the action force, F_A, is equal to 625 N, that is, one fourth of the weight of the suspended object. Remember that the fixed pulleys just change the direction of the force.

The configuration of Fig. 5.11 can accommodate a larger number of pulleys, since the number of fixed pulleys is the same as the number of pulleys that supports the movable load. Thus, for three pulleys in support, there will be six upward vertical forces, establishing the equilibrium of the set, and the mechanical advantage will be equal to six. An improvement of the configuration of Fig. 5.11 is to place the pulleys on the same axis.

In relation to the system of pulleys that guarantee the mechanical advantage, one last observation in relation to the work performed by the involved forces must be made. If the friction of the system is neglected, the work performed by the action force, F_A, by using the machine must be the same as the work that is done without it, when the force F_R is used directly. That means that to lift an object weighing W

Fig. 5.11 Combination of
two fixed pulleys with two
suspended movable pulleys,
each one with an even
number of forces

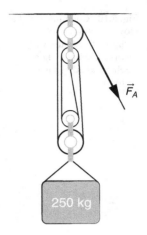

from the ground up to height h with a system of pulleys with mechanical advantage two, $(F_A = W/2)$, we have to pull a cable with length d, which will be twice the height, $d = 2\,h$. Thus, the work:

$$\tau = F_A d = Wh.$$

5.6.2   Traction Systems

In different health treatment modalities, mainly that of orthopedics related to fractures, it will be necessary to immobilize the affected region, to ease its weight, and even exert traction during the treatment. For this, a system of pulleys can be employed. The following examples exploit some of these systems.

Example 5.1 Find the magnitude and direction of the traction force exerted on the patient's leg represented in the figure of Example 5.1. The two pulleys are in the same plane.

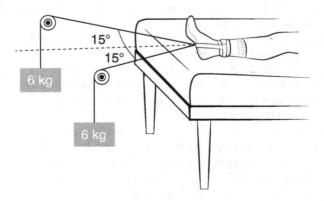

In this case, both pulleys are fixed and, hence, just change the direction of the forces. Thus, it is possible to exert forces of traction of the same value as the suspended weights, each one with 60 N. The resultant force of traction will be horizontal, directed to the left whose magnitude is given by the law of cosines.

$$T = \sqrt{(60\,\text{N})^2 + (60\,\text{N})^2 + 2(60\,\text{N})(60\,\text{N})\cos 30°} = 115.9\,\text{N}.$$

Example 5.2 An equivalent arrangement of Example 5.1 is shown in the figure of Example 5.2. Here the pulleys are placed in different positions and a single cable is utilized with one extremity attached to the wall and the other sustaining the weight for the traction.

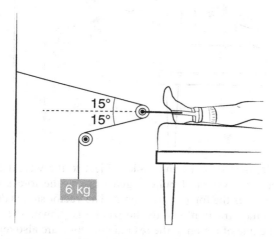

In this configuration a single weight is used to exert two forces of the same magnitude of 60 N each, whose resultant is 115.9 N, that is, greater than the load itself.

Example 5.3 Analyze now a system of pulleys, shown in the figure of Example 5.3, used for traction of an immobilized femur.

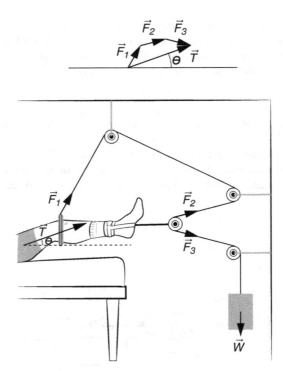

In this example, there is only one cable subject to the weight force W, of the suspended object. The system of pulleys gives rise to the forces F_1, F_2, and F_3 whose vector sum gives the force of traction T. The vector sum is drawn where the angle θ of the resultant traction with the horizontal is shown. The same angle and the direction of the line of action of the resultant traction are also represented in the figure of the system of traction.

Exercise 5.13 Find the magnitude and the direction of the force of traction of the system of Example 5.3 for the case in which the magnitude of forces obeys the following specifications: the magnitudes of vectors F_1, F_2, and F_3 are equal and the value is 45 N.

Example 5.4 Determine the intensity and the direction of the traction force in the fixed pulley attached to the patient under traction, shown in the figure of Example 5.4.
$\sin 45° = \cos 45° = 0.707$.

The intensity of each of the three forces of traction is $= 50$ N. Applying the condition of equilibrium relative to translation:

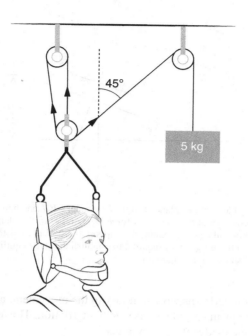

Vertical: $T_Y = 50\text{ N} + 50\text{ N} + (50\text{ N})\sin 45° = 135.4\text{ N}$
Horizontal: $T_X = (50\text{ N})\cos 45° = 35.4\text{ N}$
$T^2 = T_Y^2 + T_X^2$; hence, $T = 139.9\text{ N}$
$\sin\theta = T_Y/T = 0.967$. Hence, the angle of T with the horizontal is $\theta = 75.3°$.

5.7 Inclined Plane

In Example 1.6 of Chap. 1, we analyzed the forces acting on a child on an inclined plane. There we could see that the force necessary to maintain a child on a plane in static equilibrium corresponds to a fraction of the weight of this child. More exactly, this force of action is equal to the weight of the child, multiplied by the sine of the angle of the inclination of the plane. Analyze Fig. 5.12.

To transport an object from bottom to top using an inclined plane of height h, the action force to be employed will be smaller, as the angle θ of inclination is smaller, but, in contrast, the length d of the ramp will be longer. This occurs because the work done with or without a machine must be the same, if friction is neglected.

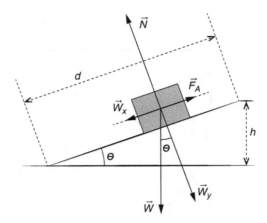

Fig. 5.12 Illustration of an inclined plane at angle θ, ramp length d, and height h. The weight force W of the object, the component of the weight force W_X parallel to the plane, the component W_Y of weight force perpendicular to the plane, the support force N normal to the surface, and the magnitude of the action force $F_A = W_X$ required to establish the static equilibrium of the object on the inclined plane are shown in the diagram

The force of friction performs work contrary to the movement, and the action force has to be larger than in the ideal case without friction. However, this does not reduce the mechanical advantage very much.

Analyzing the diagram of forces in Fig. 5.12, we can establish for the situation without friction that:

- The magnitude of the action force is given by: $F_A = W_x = W \sin \theta$
- As $\sin \theta = \frac{h}{d}$, we can write: $F_A = W \frac{h}{d}$
- Thus, the mechanical advantage is given by: $MA = \frac{F_R}{F_A} = \frac{Wd}{Wh} = \frac{d}{h}$
- The expression for the mechanical advantage can be also obtained by using the fact that the work done by the resistance force is equal to the work done by the action force, without friction: $\tau = F_R h = F_A d$

Exercise 5.14 In hospitals, in the street and in public places, it is important to have access ramps for physically disabled persons so that they may move in wheel chairs by themselves or with help. In such a situation find the force necessary to maintain a person in a wheelchair on a ramp one meter long with a rise of 0.2 m when the total weight of the person and the wheelchair is 800 N (neglect friction). What is the mechanical advantage of this inclined plane?

Exercise 5.15 The figures (a), (b), and (c) of exercise 5.15 represent situations in which simple machines based on pulleys, wheels or inclined planes are employed. Determine for each figure the action force and the mechanical advantage of the system.

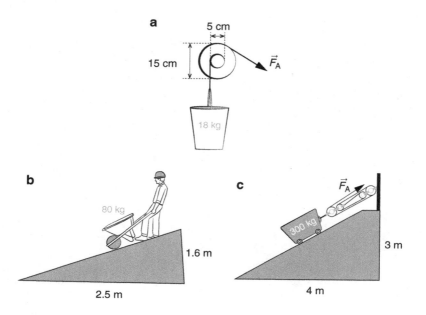

5.8   Answers to Exercises

Exercise 5.1–5.4 Representation of levers and diagrams of corresponding forces.

Exercise 5.5 $F_A = 400$ N; third class lever.

Exercise 5.6 $F_R = 187.5$ N; second class lever.

Exercise 5.7 $F_R = 1{,}500$ N; first class lever.

Exercise 5.8 For biceps—third class lever; for triceps—first class lever.

Exercise 5.9 The torque is maximum when the angle between the arm and forearm is 90°.

Exercise 5.10 $F = 491.7$ N; third class lever.

Exercise 5.11 $F = 506$ N; first class lever.

Exercise 5.12 $F_{\text{BICEPS}} = 67$ N; $F_{\text{REACTION}} = 587$ N.

Exercise 5.13 $T = 120$ N.

Exercise 5.14 $F_A = 160$ N; MA = 5.

Exercise 5.15 (a) $F_A = 60$ N; MA = 3; (b) $F_A = 431$ N; MA = 1.86; (c) $F_A = 360$ N; MA = 8.33.

Chapter 6
Muscle Force

Incorrect posture adopted daily causes immense forces on the column that are, in general, responsible for low back pain. The forces that act on the hips during a walk are very large. Simplified models are used to calculate the magnitude of the involved forces. The principles that orient these calculations are the conditions for the static equilibrium of the body or parts of it.

6.1 Objectives

- To establish and to apply the equilibrium conditions of a rigid body
- To find the forces applied on the hips during a walk
- To calculate the forces applied to the spine and to verify the influence of posture on lifting a heavy object from the floor

6.2 Equilibrium Conditions of a Rigid Body

The equilibrium of a body can be static or dynamic. In this chapter let us discuss the static equilibrium conditions that must guarantee that neither translation nor rotation of a body occurs. The conditions for equilibrium were already discussed separately in Chaps. 1 and 2.

1. Necessary condition for a body to not undergo translation—the vector sum of all the forces applied to the body must be zero, that is, the resultant force acting on the body must be nil:

E. Okuno and L. Fratin, *Biomechanics of the Human Body*, Undergraduate
Lecture Notes in Physics, DOI 10.1007/978-1-4614-8576-6_6,
© Springer Science+Business Media New York 2014

$$\vec{R} = \vec{F}_1 + \vec{F}_2 + \vec{F}_3 + \cdots + \vec{F}_n = \sum_{i=1}^{i=n} \vec{F}_i = 0, \qquad (6.1)$$

which is equivalent to stating that their orthogonal components must be zero:

$$\vec{R}_x = \sum \vec{F}_x = 0 \quad \text{and} \quad \vec{R}_y = \sum \vec{F}_y = 0. \qquad (6.2)$$

This condition, however, does not impede the rotation of an object, as in the case of a couple, presented in Chap. 2. Hence, it is necessary, but not sufficient for an object to be in equilibrium.

2. Sufficient condition for a body to not undergo rotation—the sum of all the torques exerted on the body must be zero:

$$M_T = M_1 + M_2 + M_3 + \cdots + M_n = \sum_{i=1}^{i=n} M_i = 0. \qquad (6.3)$$

Remember that all of the torques must be calculated about the same point (fulcrum).

The principle of rotational equilibrium requires, furthermore, that the imaginary line passing through the center of gravity must pass through the support area, as already stated in Chap. 3.

6.3   System of Parallel Forces

A system of parallel forces occurs when all of the forces acting on a rigid object are applied perpendicularly to a given segment of a straight line. See the illustration in the figure of Example 6.1. In this case, the conditions for static equilibrium for the forces and their respective torques provide algebraic equations from which the unknowns can be obtained by calculation. The solution of the problem can begin with the equation of torques. Let us choose as the rotation axis, for convenience, the point of application of an unknown force, since its torque is eliminated from the equation because the lever arm of this force will be zero. Then, use the fact that the resultant of all of the forces applied must be zero or another equation of torque can be written, choosing a new axis of rotation.

Example 6.1 Alice and Paul carry an object with 50 kg mass which rests on a 3.0 m board, whose weight of $W_2 = 100$ N is applied at its center of gravity, as shown in the figure of Example 6.1. The object is at 1.0 m from Paul's end of the board. Determine the magnitudes of the forces that Alice (F_A) and Paul (F_P) must exert to support this load.

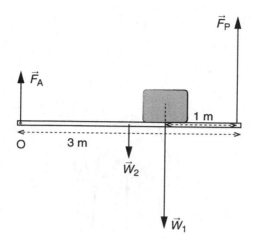

Unknowns: F_A and F_P. Let us choose the point O as the fulcrum. The feature that makes this point convenient is the null lever arm of the force F_A and hence its torque is zero about O. Thus, applying the second condition of equilibrium:

$M_T = - W_2(1.5 \text{ m}) - W_1(2.0 \text{ m}) + F_P(3.0 \text{ m}) = 0,$
$M_T = 0 = - (100 \text{ N})(1.5 \text{ m}) - (50 \text{ kg})(10 \text{ m/s}^2)(2.0 \text{ m}) + F_P(3.0 \text{ m}),$
$F_P = [(150 + 1,000)\text{N m}]/3.0 \text{ m} = 383.3 \text{ N},$
$F_P = 383.3 \text{ N}.$

Applying now the first condition of equilibrium that the resultant of forces must be zero:

$F_A + F_P = W_1 + W_2,$
$F_A = W_1 + W_2 - F_P = 500 \text{ N} + 100 \text{ N} - 383.3 \text{ N} = 216.7 \text{ N},$
$F_A = 216.7 \text{ N}.$

In Example 6.1, all of the forces that act on the board are applied on the same segment of the straight line, that is, the board.

In Example 6.2 and Exercises 6.1 and 6.2, all of the forces that act on the body are parallel, perpendicular to the ground, but applied on different levels in relation to the vertical axis. To solve these exercises, all of the force vectors can be transported to the same level, for example, to the ground, as already seen in Chap. 1, in vector addition.

Example 6.2 Find the magnitude of the contact forces exerted by the ground on the right foot N_r and on the left foot N_l of a man standing erect, shown in the figure of Example 6.2. The mass of this man is 70 kg and the location of the center of gravity is indicated in the figure. His feet are 30 cm apart and the line that passes through C.G. passes midway between his feet.

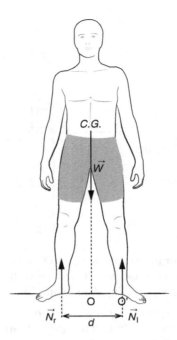

Let us begin by transporting the weight force to the level of the ground. We choose O' as the axis of rotation, at the left foot, for convenience, and apply the second condition of equilibrium:

$$M_T = (700\ \text{N})(0.15\ \text{m}) - N_r(0.30\ \text{m}) = 0,$$
$$N_r = (105\ \text{N m})/0.30\ \text{m} = 350\ \text{N}.$$

By using the second condition of equilibrium we obtain:

$$N_r + N_l = W = 700\ \text{N},$$

Hence, as $N_r = 350\ \text{N}$, $N_l = 350\ \text{N}$.

Exercise 6.1 Consider a man with 70 kg mass, who injured his left foot, as illustrated in the figure of Exercise 6.1. As a consequence, the contact force exerted by the ground on his left foot is changed to 200 N. His feet are 30 cm apart. Find the distance of the injured foot to the line that passes through the C.G. Calculate also the contact force exerted by the ground on the good right foot.

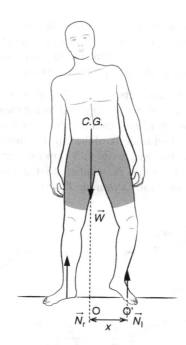

Exercise 6.2 An athlete is doing push-ups on the ground, as can be seen in the figure of Exercise 6.2. The mass of this athlete is 60 kg and the line of action of the weight force passing through the body's center of gravity is located on the ground at 1.0 m from the toes and at 0.6 m from the hands. Find the contact forces exerted by the floor on the toes and hands, for the posture shown in the figure.

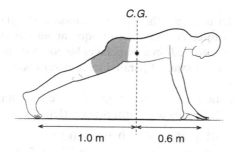

6.4 System of Nonparallel Forces

The forces that occur inside or on the body are not always parallel, but form an angle between them. One way to obtain an exact solution of such situations is to decompose the forces in some convenient orthogonal directions. An interesting way to choose the x and y axes, in order to decompose the forces, is that in which one of the components of the torque becomes zero, not causing rotation, although it can cause compression or tension on an articulation. The other component will be perpendicular to the lever arm and will produce rotation.

Example 6.3 Consider a horizontally outstretched arm, as shown in the figure of Example 6.3. In this posture, the arm is supported by the muscle force F, exerted by the deltoid muscle attached at 12 cm from the shoulder articulation at an angle of 15° to the humerus. The weight force W of the arm-forearm-hand with a magnitude of 44 N is exerted at the center of gravity at 30 cm from the shoulder articulation. There is a third force, of contact N, applied to the humerus at the shoulder articulation to equilibrate the arm. Determine F and N.

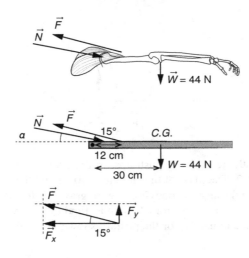

For convenience let us locate the axis of rotation at the shoulder articulation, because the unknown force N exerts zero torque about this point. We begin by decomposing the muscle force F in a perpendicular component, F_y, to the arm and the other parallel, F_x. The torque of parallel force is zero because its line of action passes through the axis.

We apply the condition that the net torque must be zero, remembering that the torque due to N is null, since its lever arm is zero. Hence,

$$W \times (0.30 \text{ m}) = F \times 0.259 \times 0.12 \text{ m},$$
$$F = (44 \text{ N})(0.30 \text{ m})/0.259(0.12 \text{ m}); \ F = 425 \text{ N}.$$

Now we apply the condition that the resultant of all applied forces must be zero:

$$W + F + N = 0.$$

We have to decompose the forces into vertical and horizontal components:

$$F_y = (425 \text{ N}) \times \sin 15° = (425 \text{ N}) \times 0.259 = 110.0 \text{ N},$$
$$F_x = (425 \text{ N}) \times \cos 15° = (425 \text{ N}) \times 0.966 = 410.5 \text{ N},$$

$$N_y = N\sin \alpha,$$
$$N_x = N\cos \alpha.$$

The resultant of the vertical components of forces must be zero:

-44 N $+ 110.0$ N $- N_y = 0$; from this equation we obtain:

$$N_y = 66.0 \text{ N.}$$

The resultant of the horizontal components of forces must be zero:

-410.5 N $+ N_x = 0$; hence,
$N_x = 410.5$ N.

Using the Pythagorean theorem, we have:

$$N_x^2 + N_y^2 = N^2 = (410.5 \text{ N})^2 + (66.0 \text{ N})^2 = 172{,}866.25 \text{ N}^2;$$

Hence, $N = 416$ N.
Now we have to find the angle α.
Knowing that $N\sin \alpha = 66$ N, we obtain:
$\sin \alpha = 66.0$ N$/416$ N; $\alpha = $ arc sin $0.159 = 9.1°$.

Exercise 6.3 A person with 100 kg mass is standing erect on the tip of both feet, as shown in the figure of Exercise 6.3. Note that each foot is subject to the normal force, equal to half of his or her weight. Find the contact force C exerted by the bones, tibia and fibula, on the ankle joint, and the muscle force F on the tendon of calcaneus at 37° with the horizontal. In this posture, the distance between the application points of the contact force and the muscle force in a longitudinal cut is 5 cm, and between the application points of the contact force and the normal force, it is 6 cm.

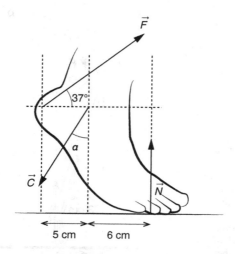

Suggestion: consider the point of application of the force C as the axis of rotation.

6.5   Forces on the Hip

When we walk, we are briefly standing erect on one leg, which changes at every footstep. At this moment, the center of gravity must lie on an imaginary line passing simultaneously through the vectors weight force W of the body and normal force N, which equilibrates W. N is exerted by the floor, as seen in Chap. 3 and is located on the foot that touches the ground. At each footstep, the articulation of the hips (acetabulum of the femur that is of the ball-socket type) exerts on the head of the femur of the leg that supports the weight, a contact force C larger than twice the body weight force. The hip abductor muscle force F is exerted by the muscles gluteus medius, gluteus minimus, and *tensor of fascia lata femoris* on the great trochanter of the femur. V.T. Inman wrote in 1947 the paper *Functional Aspects of the Abductor Muscles of the Hip* in the J. Bone Joint Surg. 29 607–619. The results of the measurements of forces exerted by the abductor muscles were presented in this paper. He obtained the value of 70° for the angle of the abductor force to the horizontal and the average distances involved in the leg of an adult, as illustrated in Fig. 6.1a. Figure 6.1b shows a simplified geometric model with the applied forces on the inferior right limb with the foot on the floor. Thus, when we walk, the magnitude of the contact force C changes drastically, according to the body weight sustained by one foot or the other alternately.

When a person walks, the line of action of the weight force that passes through the center of gravity is placed instinctively on the foot which is on the floor. When we walk normally, this process occurs automatically, but the equilibrium becomes disastrous when we try to walk very slowly, mainly with our eyes closed. Our body

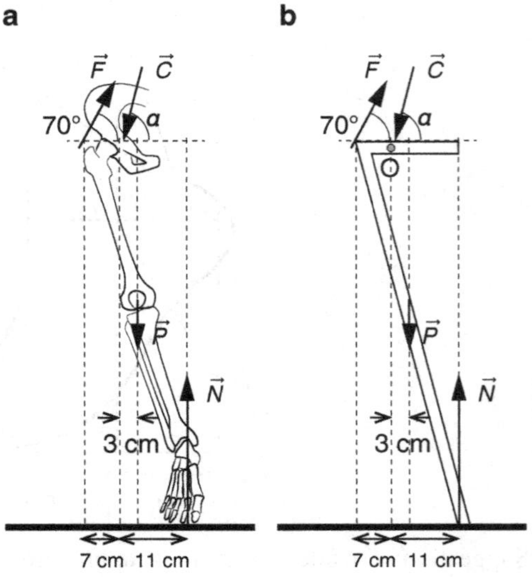

Fig. 6.1 (a) An adult standing on his right foot, in static equilibrium. The forces and the distances involved are indicated. P is the weight of the set thigh-leg-foot and N, the normal reaction to the weight force W of the body. (b) Sketch of the geometric model of the leg of (a)

oscillates forward and backward, seeming to not have decided where to put the line of action of the weight force which is along the center of gravity, whether on the right foot or on the left foot.

We are talking about walking which deals with dynamics, but the forces involved can be analyzed, considering that the body is in static equilibrium momentarily, while one of the feet is on the ground.

Example 6.4 Consider a person with 60 kg mass standing erect on her or his right foot. The mass of each set thigh-leg-foot is of 9.0 kg. Use Fig. 6.1b and determine the intensities: (a) of the hip abductor muscle force F, exerted by the gluteus muscle which makes an angle of 70° to the horizontal on the great trochanter of the femur, and (b) of the contact (reaction) force C, exerted by the acetabulum (socket of the pelvis) on the femur head, as well as the direction of the contact force in relation to the horizontal.

We locate the rotation axis O, where the contact force C is applied. Next we decompose the forces F and C in their orthogonal components:

$F_y = F\sin 70° = F \times 0.940$ and $F_x = F\cos 70° = F \times 0.342$,
$C_y = C\sin \alpha$ and $C_x = C\cos \alpha$.

Now we apply the condition that the net torque is zero about the axis of rotation at O:

$0 = - F_y(0.07 \text{ m}) - W(0.03 \text{ m}) + N(0.11 \text{ m})$,
$0 = - F \times 0.940 \times (0.07 \text{ m}) - (90 \text{ N})(0.03 \text{ m}) + (600 \text{ N})(0.11 \text{ m})$; from this equation we obtain:

$$F = 962 \text{ N}.$$

Hence, $F_y = 904.3 \text{ N}$ and $F_x = 329.0 \text{ N}$.

Now we apply the condition that the sum of y and x components of forces must be equal to zero:

$0 = F_y - C_y - P + N = (904.3 \text{ N}) - C_y - (90 \text{ N}) + (600 \text{ N})$,
$C_y = 1{,}414.3 \text{ N}$,
$0 = F_x - C_x$,
$C_x = 329.0 \text{ N}$.

Then, doing the vector sum: $C_y{}^2 + C_x{}^2 = C^2 = (1{,}414.3 \text{ N})^2 + (329.0 \text{ N})^2$
$C = 1{,}452 \text{ N}$; as expected, C is greater than twice the body weight of this person. Knowing that $(1{,}452 \text{ N})\cos \alpha = 329.0$, we obtain: $\alpha = \text{arc cos } 0.226$; $\alpha = 76.9°$.

Exercise 6.4 Repeat Example 6.4, assuming that the mass of the person has doubled, i.e., 120 kg. The mass of the set thigh-leg-foot is now 20.5 kg. All the other values are the same. Discover the consequences of this overweight.

Analyzing the results of Example 6.4 and of Exercise 6.4, we observe that the relation between the intensities of each of the forces F and C and the body weight force W is maintained constant. That is: $F/W = 1.6$ and $C/W = 2.4$.

6.6   Forces on the Spinal Column

After the age of 40, rare are the persons who have no lower back pain or lumbar region pain caused by incorrect posture during their life. Many persons may come to have lumbar region pain, due to incorrect lifting of weight, for example, lifting a child from the ground. To understand a little about the origin of this pain, let us analyze the forces involved in such cases.

A human spinal column is divided in four parts, from top to bottom: the cervical, made up by seven vertebrae; the thoracic, by twelve vertebrae; the lumbar, by five fused vertebrae in the sacrum; and four fused vertebrae in the coccyx. As already discussed in Chap. 1, the vertebrae of the human spinal column increase in size continuously from top to bottom to bear increasingly heavy weight. Between the vertebrae there are intervertebral discs made of fibrous material with the purpose of damping the forces and the impacts suffered by the spinal column, for example, in footraces, when, briefly, the person is in the air and lands on one foot, differently when the person walks and never leaves the ground. The average spinal column length of a normal adult is 70 cm, which at the end of the day can be shortened by 1.5 cm, but is recovered after one night in a horizontal position. This fact is easily observed by the driver when he or she notes that the adjustment of the rear view mirror inside the car differs between the morning and at the end of a day. As the years pass, unhappily, the spinal column is shortened little by little, due mainly to osteoporosis.

The principal muscles used to bend the back or to lift objects from the ground are spinal erector muscles. They connect the ileum and the lower part of the sacrum to all of the lumbar vertebrae and to four thoracic vertebrae. Research performed by L.A. Strait, V.T. Inman and H.J. Ralston and published in the Amer. J. Phys. 15, 377–378 (1947) has shown that during a back flexion, the forces exerted by the spinal erector muscles on the spinal column can be represented by a single muscle force. The spinal column can be considered a rigid body, and the force is applied at a point at 2/3 of its length from the sacrum, and forming an angle of 12° with the column. The axis of rotation is located at the fifth lumbar vertebra. The contact (reaction) compressive force of largest intensity is applied exactly there during the bending of the back. If this force surpasses a certain limit, the intervertebral disc flattens and its diameter increases, pressing on the nerve and, as a consequence, causing lumbar pain.

6.6.1   Forces Involved in the Spinal Column
When the Posture Is Incorrect

One of the incorrect postures is that in which the spinal column is curved (without bending of the knees) whether to brush the teeth over a washbasin or to lift a heavy object from the floor. For these situations, it is truly impossible not to curve the spine, if the knees are not bent. Even in the case of brushing the teeth over a low washbasin in which no weight is being lifted, we verify that when the column is just curved, huge muscle forces are involved. These forces are stronger the more the column is bent, which happens with taller persons, since the height of the washbasin is not personalized, but standardized. Besides, the height of everything that surrounds us is standardized and can be the source of low back pain as is the case of the height of desks, of chairs, of kitchen sinks or of toilet seats from which tall and old persons have extreme difficulty in getting up.

Nowadays, many persons are learning to behave correctly, with posture reeducation, in order to avoid pain that results from incorrect posture.

In Example 6.5 we will calculate the value of the forces involved in a spinal column during its flexion, without bending the knees, using a simplified model. The obtained results show that they are very reasonable, in spite of the simplified model, compared to the complexity of the human body.

Example 6.5 Consider the situation of a person bending over to lift a heavy object from the floor, with straight legs, i.e., without bending of the knees, as shown in the Fig. 6.2 The column, considered as a rigid body, makes an angle Φ of 30° to the horizontal. The weight W_1 of the trunk of this person is of 300 N and is applied at the middle of the column. The weight W_2 of the head + two (arms/forearms/hands) of 150 N added to the weight of an object of 200 N acts on the upper part of the column. The muscle force F exerted by the erector muscles acts at 2/3 of the column length that is 70 cm, forming an angle of 12° with the column. The contact force C compresses the intervertebral disc between the sacrum and the fifth lumbar vertebra. Assuming that the body is in equilibrium, determine the intensities of F, C and the direction of C. The given data refer to a standard Caucasian adult of 70 kg.

We begin solving the problem by decomposing all of the forces in their components orthogonal to the bar that simulates the column. As the column makes an angle of 30° with the horizontal, and the sum of internal angles of a triangle is 180°, the angle of the weight forces W_1 and W_2 with the bar is 60°. We also know that:

$\cos 60° = 0.500$; $\sin 60° = 0.866$; $\cos 12° = 0.978$; $\sin 12° = 0.208$,
$W_{1y} = W_1 \sin 60°$; $W_{1x} = W_1 \cos 60°$; $W_{2y} = W_2 \sin 60°$; $W_{2x} = W_2 \cos 60°$,
$F_y = F \sin 12°$; $F_x = F \cos 12°$,
$C_y = C \sin \alpha$; $C_x = C \cos \alpha$.

We apply the condition of total torque equal to zero about the axis of rotation, conveniently chosen to be in the fifth lumbar vertebra, since the torque due to the contact force C which is unknown becomes zero. We remember that only the perpendicular component produces rotation:

$$-W_{1y}(0.35 \text{ m}) - W_{2y}(0.70 \text{ m}) + F_y(0.467 \text{ m}) = 0.$$

The only unknown term in the above equation is $F_y = F \sin 12°$.
Hence, $F = 3{,}123.6$ N.

Now we apply the condition that the resultant of all of the applied forces on the column in static equilibrium is equal to zero. Hence, the resultant of their orthogonal components must also be zero.

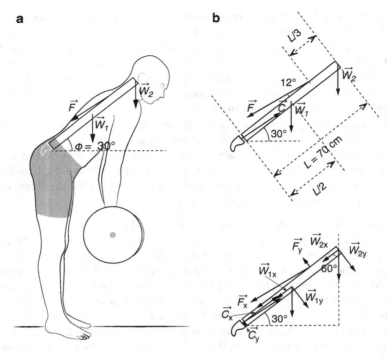

Fig. 6.2 (a) An adult lifting a weight from the floor. The spinal column is represented by a *bar*. (b) Simplified geometric model, with all of the applied forces on the column and the distances involved

$0 = -W_{1y} - W_{2y} - C_y + F_y$; The only unknown is $C_y = 86.8$ N,
$0 = -W_{2x} - W_{1x} - F_x + C_x$; The only unknown is $C_x = 3,380.0$ N.
Performing the vector sum: $C_y^2 + C_x^2 = C^2$ and extracting the square root, we obtain:

$$C = 3,381.0 \text{ N.}$$

This result informs us that if a person with a weight force of 700 N, which is reasonable for a person with a trunk weighing 300 N, lifts an object with 200 N weight incorrectly, the intervertebral disc between the fifth lumbar vertebra and the sacrum is subject to a compression force somewhat smaller than five times his body weight; now the angle α can be obtained:

$$\sin \alpha = C_y/C; \ \alpha = \text{arc sin } 0.026 = 1.47°.$$

Exercise 6.5 Repeat example 6.5 considering that the person will lift an object from the floor weighing 500 N. Determine the magnitudes of muscle force *F* and that of contact force (reaction) *C* and the direction of *C*.

Now we determine for the same person of Example 6.5 the relation between the mass of an object that is lifted incorrectly, without bending of the knees and the intensity of the muscle force, *F*, and that of the contact force, *C*, applying the

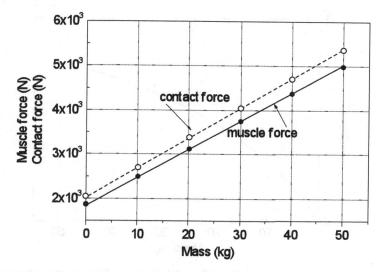

Fig. 6.3 Relation between the mass that a person lifts from the floor with both hands, incorrectly, without bending the knees and curving the column 30° to the horizontal, and the force of the spinal erector muscle and the contact compressive force

conditions for the equilibrium. The weights W_1 and W_2 considered are of a standard Caucasian adult of 1.70 m height and 70 kg, and the column length of 0.70 m. Figure 6.3 shows the obtained results.

From Fig. 6.3 we verify that even if any object is lifted (W_2 corresponds to the weight of the head plus two arms/forearms/hands), the simple fact of bending the column at an angle 30° with the horizontal, requires the muscle to exert a force of 1,874 N, which is equivalent to supporting a weight 2.6 times larger than his own body weight. This force increases linearly with the *mass* lifted, obeying an equation of a straight line:

Muscle force F (N) = 1,874.2 + 62.5 × *mass* (kg).

To lift a mass of 50 kg, the force exerted by the spinal erector muscles reaches the spectacular value of 4,999 N which is equivalent to more than 7.1 times the body weight.

In the case of the contact force C, its intensity is always somewhat greater than that of the muscle force and can be calculated for this person in this situation by the equation of a straight line:

Contact force C (N) = 2,058 + 66.2 × *mass* (kg).

The angle α of the contact force C with the column varies little with the mass that a person lifts; it is very small and the force is downward, as can be seen in Fig. 6.4. The above equations are obtained by solving Example 6.5.

Exercise 6.6 Discuss what happens when the person of Example 6.5 adopts the correct posture bending the knees to lift an object from the ground. Try to crouch correctly, bending the knees to see what changes. Solve the exercise quantitatively, considering that the angle of the column with the horizontal is 70°, and observe what changes.

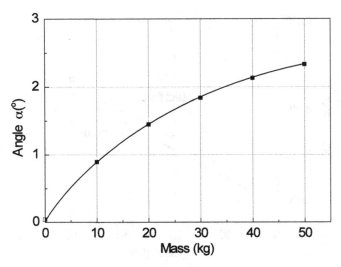

Fig. 6.4 Angle α of the contact force C with the vertebral column. The angle of this force with the horizontal can be obtained by subtracting it from 30° which is the angle of the column with the horizontal

6.6.2 Forces Involved in the Spinal Column When the Posture Is Correct

Let us now analyze what, and by how much, changes when a person lifts an object from the ground correctly, bending the knees. You probably noticed that the changes are in the angle Φ of the column with the horizontal, which becomes larger. We perform the calculations in order to obtain the intensities of muscle force F and that of contact force C, varying the angle Φ from 10° to 70° in steps of 10°. Figures 6.5 and 6.6 show the muscle forces and the contact forces as a function of angle Φ, with the mass that the person lifts of 0 kg (this person will not lift any additional weight, but will have to raise his head + two(arms/forearms/hands)), 20 kg, and 50 kg, as the parameter.

The force exerted by the spinal erector muscle decreases with the increase of the angle Φ of the column with the horizontal, which results from bending the knees and crouching to lift an object from the ground. As the column is maintained straighter (vertical), the greater is the angle Φ leading to a correspondingly smaller muscle force. The decrease is of 2.9 times if we compare the situations in which the bending of the spine in relation to the horizontal goes from 10° to 70°, independently of the lifted mass.

The decrease of the muscle force is not linear with the increase of the angle Φ. As Φ goes from 30° to 70°, the decrease obtained is 2.5 times. We remember that the curves of Fig. 6.3 were obtained for the angle Φ equal to 30° whose value is a very common incorrect posture adopted daily. These calculations show quantitatively the importance in adopting the correct posture.

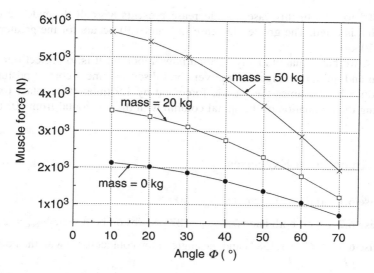

Fig. 6.5 Force exerted by the spinal erector muscle on the vertebral column as a function of angle Φ of the column with the horizontal, with the mass that is lifted as the parameter

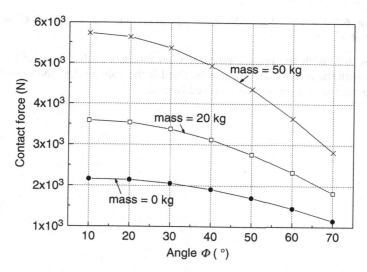

Fig. 6.6 Contact force exerted by the sacrum on the last intervertebral lumbar disc as a function of angle Φ of the column with the horizontal, with the lifted mass as the parameter

Figure 6.6 also shows the decrease of the contact compression force in the last intervertebral lumbar disc with the increase of the angle Φ of the spine to the horizontal. Calculating the decrease as a function of the angle Φ, we obtain the value of 1.9–2.0 when the bending of the column goes from 10° to 70° and of 1.8 to 1.9, from 30° to 70°. In this case, the decrease presents a small dependence on the mass that is lifted. The greater decrease (2.0 and 1.9) occurs for the greater lifted mass (50 kg).

Thus, we have quantitatively verified how important it is to protect our spinal column and, to not damage the intervertebral discs, assume a correct posture that can reduce by 2.5 times the muscle force and by 1.9 times the contact force by changing the inclination of the spinal column with the horizontal from 30° to 70°.

6.7 Answers to Exercises

Exercise 6.1 (a) $d = 21.4$ cm; (b) $N = 500$ N.

Exercise 6.2 (a) on both hands, $N_{hand} = 375$ N; (b) on both feet, $N_{feet} = 225$ N.

Exercise 6.3 (a) $C = 1,358$ N and the angle of the contact force with the horizontal is 54°; (b) $F = 997$ N.

Exercise 6.4 (a) $F = 1,913$ N; (b) $C = 2,869$ N and the angle of the contact force C with the horizontal $= 76.8°$. Note that doubling the body mass, the contact force C also has doubled.

Exercise 6.5 (a) $F = 4,998$ N; $C = 5,367$ N and the angle of C with the horizontal $= 2.3°$.

Exercise 6.6 With the correct posture, what changes is the angle of the spinal column with the horizontal. The less inclined is the column, i.e., the more erect is the column, the smaller are the involved forces.

Chapter 7
Bones

The bones that form the skeleton are made of an organic matrix—collagen, and inorganic crystals of calcium and phosphate, plus water. The bones are made to support the body's weight as well as quite larger stresses. Collagen is responsible for the bone's great elasticity and the inorganic crystals for their resistance to tensile and compressive forces. The elastic properties of the bone are compared with those of steel-reinforced concrete.

7.1 Objectives

- To introduce skeleton and bones
- To introduce the elastic properties of solids
- To discuss the concepts of elastic moduli of solids
- To discuss the elastic properties of bones
- To analyze stress on intervertebral discs
- To analyze the situations in which bones fracture

7.2 Skeleton and Bones

The skeleton is a mechanical structure and is composed of bones, cartilage, and articulations. Its main functions are: body support, movement, protection of important organs such as the brain, eyes, internal ears, heart, lungs, etc. and the storage of chemical substances, being calcium the most important. An adult skeleton is made up of 206 bones of different shapes and sizes, half of which belongs to hands and feet. Due to their composition, bones are extraordinarily strong although also extraordinarily light, mainly because of their porosity. In a standard adult the mass of all of the bones corresponds to 14–18 % of the total body mass.

E. Okuno and L. Fratin, *Biomechanics of the Human Body*, Undergraduate
Lecture Notes in Physics, DOI 10.1007/978-1-4614-8576-6_7,
© Springer Science+Business Media New York 2014

7.2.1 Composition of Bones

The bones are composed of a rigid organic matrix reinforced by the deposit of inorganic crystals that constitute the mineral bone, plus water. A standard compact bone contains in weight about 40 % of organic matrix and 60 % of mineral bone. The main component of the organic matrix is the collagen that is responsible for the great elasticity of the bone. This collagen is not the same as the collagen found in other parts of the body such as the skin. Inorganic crystals are chiefly calcium and phosphate salts, known as hydroxyapatite, whose chemical form is $Ca_{10}(PO_4)_6(OH)_2$.

Bone structure can be compared with that of concrete reinforced with steel rods. The combination of collagen fiber with hydroxyapatite crystals allows the bones to support compressive forces larger than those of concrete, allied with great elasticity.

Bone is a living tissue and undergoes changes continuously. We can say that about every 7 years a new skeleton is created. The density of the bone is 1.9 g/cm^3 and does not change as time goes by. From the age of 35–40 years onward, the mass of the bones decreases slowly at a rate of 1–2 % per year. Osteoporosis is the decrease of the bone mass, with thinning of the bone causing it to become more fragile.

Either the collagen or the mineral bone can be withdrawn from the bone separately; the first by burning in an oven and the last with an acid solution, and what remains from each process will have the original shape. Therefore, when a body is cremated, the collagen is removed and what remains from the skeleton is the mineral bone that is placed in urns.

Mineral bone is preserved during millennia, and for this reason the skeleton is an important piece in anthropological, paleontological, archaeological, and evolutionary research. Its dating can be performed by detecting[1] the free radicals produced by the environmental ionizing radiation or by the method of carbon-14. Nowadays, measurements of the specific elemental contents of the bone can give information on the type of food of prehistoric human beings.

Before the discussion about how strong bones are, let us present the elastic properties of solids.

7.3 Elastic Properties of Solids

All solid bodies undergo deformation, when external forces act on them, which may or may not be visible to the unaided eye. In general, they return to their original shape when the external applied forces are removed, if the magnitude of these forces is not too large. Naturally, any object can be deformed permanently or even broken if the deformation surpasses a limit, characteristic of each material. The intrinsic elasticity of solids, in the last analysis, is due to the electric forces between atoms and molecules that compose the object.

---

[1] Technique of dating by electron paramagnetic resonance (EPR).

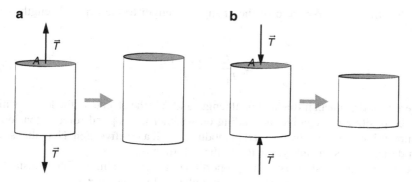

Fig. 7.1 A cylinder is subject to (**a**) tensile stress and to (**b**) compressive stress that cause elongation and shortening, respectively. Note that the change in the *shape* is exaggerated

7.3.1 Tensile Stress and Compressive Stress

Let us introduce here other forces that have not been discussed in Chap. 1. They are tension and compression exerted on the objects, as shown in Fig. 7.1. Both consist of two forces T of equal magnitude but with opposite direction that maintain the object at rest, and thus there is neither translation nor rotation, since the necessary and sufficient conditions for a body to be in equilibrium are obeyed. Tension causes an extension of the bodies and compression, on the other hand, compresses the body, shortening it slightly. Simultaneously with the elongation or the shortening, a small, in general not noticeable, narrowing or widening of the bodies occurs. If the elongation or the shortening is relatively small which is, in general, the case, we do not have to consider the modifications in its cross-sectional area.

The stress σ is defined as the ratio of the applied force called tension, T, divided by the cross-sectional area, A:

$$\sigma = \frac{T}{A}. \tag{7.1}$$

The tension, T, is the magnitude of the force, either tensile or compressive, that causes the deformation.

Hence, the unit of stress is N/m^2, the same unit as pressure, which has the special name of pascal (Pa) in the International System of Units. Let us denote tensile stress, σ_t, or compressive stress, σ_c, for the applied force either of tension or compression, respectively. In these cases the force is always applied longitudinally, i.e., perpendicularly to the area, A.

Conceptually, pressure and stress are physically equal, with pressure more used for fluids and stress for solids, chiefly by engineers who perform tests of material strength. The concept of pressure has already been presented in Chap. 1 and its use is easier and more intuitive since it is a part of our daily life. The concept of stress is more general, on the other hand, since it includes tensile stress.

The strain, ε, is the ratio of the change of length to the original length of an object:

$$\varepsilon = \frac{\Delta L}{L_i} = \frac{|L_f - L_i|}{L_i},$$ (7.2)

where L_i and L_f mean initial and final length and ΔL, the change in the length which can be positive or negative, depending on whether the applied force is tension or compression, and it is always used in modulus, with a positive sign. Note that strain is a dimensionless quantity, related to the deformation.

For very small stresses, σ is proportional to the strain, ε. The constant of proportionality is called the elasticity modulus and is given by:

$$\text{Elasticity modulus} = \text{stress/strain} = \sigma/\varepsilon.$$ (7.3)

The unit of the elasticity modulus is the same as the stress, i.e., pascal (Pa), because the strain is dimensionless.

7.4 Modulus of Elasticity

The elastic properties of solids are always described in terms of the modulus of elasticity that varies from material to material, but does not depend on the size of the object. Here we will treat two elasticity moduli of interest to us:

(a) Young's modulus that measures the resistance of solids relative to the change in their length.
(b) Shear modulus that measures the resistance to the sliding motion of different plane layers of a solid.

7.4.1 Young's Modulus Y

In this case,

$$Y = \frac{\sigma}{\varepsilon} = \frac{\frac{T}{A}}{\frac{\Delta L}{L_i}},$$ (7.4)

T is the tension applied along the cylinder as shown in Fig. 7.1 and A is the cross-sectional area of the cylinder. The larger is Young's modulus, the smaller is the elasticity of the material regarding the change in the length by tension or by compression, or in other words, it is more difficult to change its length.

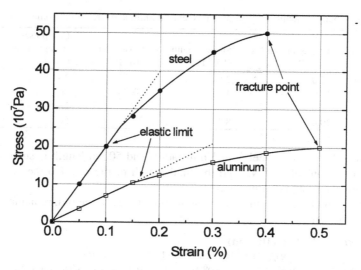

Fig. 7.2 Curves of stress against strain for steel and aluminum rods. Note that the strain is in percentage. The elastic limit and the tensile strength (fracture point) are indicated

Equation (7.4) can be written as $\sigma = Y\varepsilon$, that is an equation of a straight line, where Y is Young's modulus, the slope of the straight line. The larger is the slope of the straight line with respect to the x axis, in this case the axis of strain ε, the greater is Young's modulus. Figure 7.2 shows the graph of stress as a function of strain for steel and aluminum rods, subjected to tension. Observe that the equation of the straight line $\sigma = Y\varepsilon$ is represented by dotted lines. In the case of the steel rod, the slope of the straight line is much greater than that for the aluminum rod. We can also say that for the same applied stress, the smaller is Young's modulus, the larger is the strain produced.

The elastic phase of a material corresponds to a regime in which the object returns to its original shape, when the stress is removed; in terms of stress, it corresponds to stresses between zero and the elastic limit. Above the elastic limit, the phase is said to be plastic and the deformation produced becomes permanent until the tensile or compressive strength is exceeded, when the object breaks (fracture point). It is important to emphasize that once the plastic limit is surpassed, it will be very easy to break any piece, even made of metal, by twisting it from one side to another as we usually do when the opener of an aluminum can is removed.

For metallic objects, generally, Young's modulus has the same value for tensile or compressive stresses. This is not true for heterogeneous materials such as wood, concrete, plastic and bones, that, besides a different Young's modulus for tensile or compressive stresses, have also different elastic limits and fracture points.

Table 7.1 gives Young's modulus, the shear modulus, the elastic limit and the compressive and tensile strength for some common solids. Observe that either the elastic limit or the strength in Table 7.1 is the value of stresses the limits the elastic from the plastic phase and that causes fracture of a given solid and is measured in pascal (Pa).

Table 7.1 Young's modulus, Shear modulus, elastic limit, and strength of some common solids

Material	Young's modulus $(10^{10}$ Pa$)$	Shear modulus $(10^{10}$ Pa$)$	Elastic limit $(10^7$ Pa$)$	Compressive and tensile strength (fracture point) $(10^7$ Pa$)$
Aluminum	7.0	2.5	18	20
Copper	12.0	4.2	20	40
Granite	5.0			20
Steel	20.0	8.4	30	50

Example 7.1 A rod of steel with 1 mm radius and 50 cm length is submitted to a tension (traction) of 500 N. Find the final length of the rod. Do the same for an aluminum rod of equal dimension.

Let us begin by calculating the cross-sectional area of the rod, remembering that 1 mm $= 10^{-3}$ m.

$$A = \pi r^2 = 3.14 \times (10^{-3} \text{ m})^2$$
$$\sigma = T/A = (500 \text{ N})/(3.14)(10^{-3} \text{ m})^2 = 159.2 \times 10^6 \text{ Pa}$$
$$\varepsilon = \Delta L/L_i = \sigma/Y = 159.2 \times 10^6 \text{ Pa}/(20 \times 10^{10} \text{ Pa}) = 8.0 \times 10^{-4}$$

$\Delta L = L_i \varepsilon = (0.50 \text{ m})(8.0 \times 10^{-4}) = 4.0 \times 10^{-4} \text{ m} = 0.4$ mm that is the elongation of the steel rod. Hence, its final length is $L_f = 50.04$ cm. When the tensile stress is removed, the length of the rod returns to 50 cm.

In the case of the aluminum rod its Young's modulus is smaller than that of steel. Beforehand, we note that the elongation of this rod must be greater than that of the steel rod, since its Young's modulus is smaller.

$$\varepsilon = \sigma/Y = (159.2 \times 10^6 \text{ Pa})/(7.0 \times 10^{10} \text{ Pa}) = 22.7 \times 10^{-4}$$
$$\Delta L = (0.50 \text{ m})(22.7 \times 10^{-4}) = 11.4 \times 10^{-4} \text{ m} = 1.14 \text{ mm}.$$
Hence, $L_f = 50.11$ cm.

Exercise 7.1 The cross-sectional area of a copper rod is of 5×10^{-5} m^2. The rod is stretched to its elastic limit. (a) Find the tension (traction) applied to the rod; (b) find the strain in the copper rod; (c) determine the tension required to break this rod.

Exercise 7.2 A marble column of 2.0 m height and cross-sectional area of 25 cm^2 supports a mass of 70 tons. Young's modulus for marble is 6×10^{10} Pa and the compressive strength at the fracture point is 20×10^7 Pa. Find: (a) the radius of column; (b) the compressive stress applied to the column; (c) the shortening of the column; (d) the maximum weight that the column supports.

7.4.2   Shear Modulus S

Let us consider a block on a smooth surface. This block is subjected to a tangential force called the shear force, T, applied parallel to its upper surface while its opposite face is immovable due to the force of static friction f_s, of the same

Fig. 7.3 A block subject to shear stress that distorts the shape in a way that its cross-sectional area changes from a *square* to a *parallelogram*

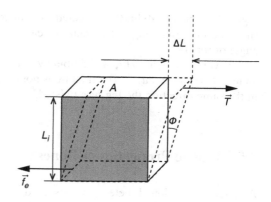

Fig. 7.4 Torque of force T applied to a cylinder fixed on the *bottom*

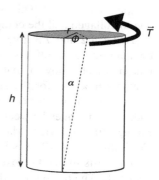

magnitude, i.e., $T = f_s$, and in the opposite direction, as shown in Fig. 7.3. The shear stress is defined as $\sigma = T/A$, where A is the area of the surface. The magnitude of the deformation is given by the shear strain, defined as $\varepsilon = \Delta L/L_i$ where ΔL is the horizontal dislocation of the surface that undergoes shear and L_i is the initial height of the block, as can be seen in Fig. 7.3.

The shear modulus, S, is given by:

$$S = \frac{\sigma}{\varepsilon} = \frac{\frac{T}{A}}{\frac{\Delta L}{L_i}}. \tag{7.5}$$

In this case, the shear strain $\varepsilon = \Delta L/L_i = \tan \phi \approx \phi$ (block inclination measured in radians). Observe that the shear force is always parallel to the area A.

The values of the shear modulus of some materials are given in Table 7.1. Generally, the value of S is between $Y/2$ and $Y/3$, meaning that it is easier to deform solids with shear stress than with tensile or compressive stress.

Another type of stress is the torsion stress that can cause a deformation similar to that produced by shear stress. In the case of shear, the force is applied parallel to the area A, while in the case of torsion, the torque is applied in the direction of rotation.

Figure 7.4 shows a cylinder subjected to a torsion stress. The angle α is called the shear deformation angle and can be calculated from $\alpha = r\phi/h$ and ϕ is the angle of torsion.

It is very common to break the tibia by twisting the leg after a misstep caused by an unnoticed irregularity. The fracture is not transversal, but in the form of a spiral, in the same direction the leg was twisted.

7.5   Elastic Properties of Bones

As already mentioned, heterogeneous solids composed of materials with different elastic characteristics present different values of Young's modulus for tensile and compressive stresses. This is the case of the bone, for which the tensile strength is due to collagen and the compressive strength, to the mineral bone.

Tables 7.2 and 7.3 give the elastic characteristics of bones under tension and compression, respectively. The bone shear modulus is not given in the table, but measurements showed that its value is about 1.0×10^{10} Pa, hence similar to Young's modulus for the femur subjected to compressive stress. Research done during more than 25 years by H. Yamada with "fresh" bones of adult Japanese cadavers with ages between 40 and 59 years provided data published in the book Strength of Biological Materials, edited by F.G. Evans, Williams and Williams Co. Baltimore, 1970. It is important to point out that the data have been obtained through tests similar to those carried out in an engineering laboratory with inorganic materials. The behavior of these bones inside the body can be very different, since they are connected to muscles through ligaments and tendons and the tests are made only with bones. It is also not known whether or not the values obtained by Yamada depend on race.

Table 7.2 Tensile properties of bones

Bone	Young's modulus (10^{10} Pa)	Tensile strength (fracture point) (10^7 Pa)	Maximum strain
Human femur	1.6	12.1	0.014
Horse femur	2.3	11.8	0.0075
Human vertebra	0.017	0.12	0.0058

Table 7.3 Compressive properties of bones and of intervertebral discs

Bone	Young's modulus (10^{10} Pa)	Compressive strength (fracture point) (10^7 Pa)	Maximum strain
Human femur	0.94	16.7	0.0185
Horse femur	0.83	14.2	0.024
Human vertebra	0.0088	0.19	0.025
Intervertebral disc	0.0011	1.10	0.30

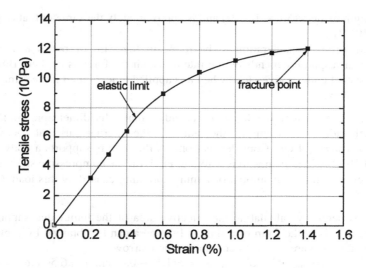

Fig. 7.5 Graph of tensile stress versus strain. Tension is applied to a human humerus. The elastic limit and the tensile strength are indicated

The forces that act on the bones are classified as longitudinal that include compression and tension and transversal that are the shear forces. Tensile forces are applied, for example, when traction is exerted on a leg in an equipment of physiotherapy and compressive forces, when a person suffers a fall from a certain height or when he or she jumps with a parachute, for example, and reaches the ground.

In accidents, involving a collision between two cars, the body of the driver goes forward instantaneously with a small inclination and strikes the steering wheel. In this movement, some vertebrae are compressed anteriorly and tensioned posteriorly which causes horizontal fracture of vertebra, according to information given by the orthopedist, Dr. Alexandre Sadao Iutaka,[2] a specialist in the spinal column.

Figure 7.5 shows the curve of the tensile stress applied to the human humerus against strain. Similar curves are obtained for the radius and the ulna, with the main difference in the value of the tensile strength.

Young's modulus of the human femur under traction is 1.7 times larger than that under compression. This means that for the same value of stress, if the applied force is of compression, the deformation produced is 1.7 times larger than that caused by tension. From the tables, we verify that the compression strength of the human femur is 1.38 times larger than the tensile strength, or in other words, the tibia or the femur is more susceptible to fracture under tension than under compression. Note that the compressive strength of the human femur (16.7×10^7 Pa) is about 1/3 of

[2] We acknowledge the information given by Dr. Iutaka, orthopedist of Hospital das Clínicas of Faculdade de Medicina of S. Paulo University.

the strength of steel (50×10^7 Pa) and is approximately the same of that of granite (20×10^7 Pa).

An important consideration to be made in the case of bones is that their mechanical response depends on the rate at which the force is applied. Bones are more resistant under a given load when it is applied quickly than if the same load is applied slowly.

Example 7.2 Consider the femur of an adult male. Its dimensions are: length = 50 cm, radius = 1.5 cm and the radius of the internal part that contains the bone marrow = 0.4 cm. Consider that one of the femurs supports a body weight force of 700 N of a person who is walking. Find: (a) the compressive stress applied to this femur and (b) the amount of femur shortening caused by this load. Consult Table 7.3.

(a) Let us begin by calculating the effective area of the femur that supports the weight: it is in the form of a ring and the area can be obtained by subtracting from the total area the area containing the marrow:

$A_{\text{effetive}} = \pi(r_{\text{femur}})^2 - \pi(r_{\text{marrow}})^2 = \pi(1.5^2 - 0.4^2 \text{ cm}^2) = 6.56 \text{ cm}^2 = 6.56 \times 10^{-4} \text{ m}^2$

$\sigma = T/A = (700 \text{ N})/(6.56 \times 10^{-4} \text{ m}^2)$.

Hence, $\sigma = 1.07 \times 10^6$ Pa

Comparing this result with atmospheric pressure, we verify that this stress is equivalent to 10.56 atm, which is very large.

(b) $\varepsilon = \sigma/Y = (1.07 \times 10^6 \text{ Pa})/(0.94 \times 10^{10} \text{ Pa}) = 1.138 \times 10^{-4}$

$\Delta L = L_i\varepsilon = (0.50 \text{ m})(1.138 \times 10^{-4}) = 0.57 \times 10^{-4} \text{ m} \approx 0.06 \text{ mm}$

Example 7.3 Studies with human bones demonstrate that they behave elastically, for deformations smaller than 0.5 %. Find the tension and the compression at the elastic limit for an adult humerus which is 0.20 m in length with a cross-sectional area of 3.0 cm^2. Suppose that the elastic properties of the humerus are the same as for the femur. Consult Tables 7.2 and 7.3.

Tension $= T_t = \sigma_t A = Y_t \varepsilon A = Y_t(\Delta L/L_i)A = (1.6 \times 10^{10} \text{ Pa})(0.005)(3.0 \times 10^{-4} \text{ m}^2)$

$T_t = 24$ kN

Compression $= T_c = \sigma_c A = Y_c \varepsilon A = Y_c(\Delta L/L_i)A = (0.94 \times 10^{10} \text{ Pa})(0.005)(3.0 \times 10^{-4} \text{ m}^2)$

$T_c = 14$ kN

Note that the humerus remains in the elastic phase for loads (masses) as large as 2.4 tons, when tensioned, and 1.4 tons, when compressed.

Exercise 7.3 In Example 1.1, the force applied to a leg of an adult in the traction apparatus was found to be 76.6 N. Supposing that the tibia has the same elastic properties as the femur, find its elongation in percentage when this force is applied. Consider the cross-sectional area of the tibia to be 3.3 cm^2.

R.W. McCalden, J.A. McGeough, M.B. Barker, and C.M. Court-Brown published in 1993 a paper with the title *Age-Related changes in the Tensile Properties of Cortical Bone* in the J. Bone and Joint Surg. Am. 75 (A-8)

1193–1205. The results obtained with 235 specimens with ages between 20 and 102 years showed that the deterioration for each decade of compact bone is 5 % in the strength and 9 % in the maximum strain. Also, Yamada (1970) has measured the deterioration of the compact bone of the femur with age. He used fresh femurs of female and male cadavers, separated in three groups with ages between 20 and 39, 40 and 59 and 60 and 89 years and measured the compression that fracture them. He obtained for the above sequence the following results: 51, 48, and 43 kN for men and 42, 40, and 35 kN for women. These results explain why older women are more likely to have their femur fractured than men.

Exercise 7.4 A hair breaks under a tension of 1.2 N. Supposing that its tensile strength is 1.96×10^8 Pa, find the radius of the hair. This hair belongs to an oriental whose cross-section is a circle, while that of a white occidental is slightly elliptical and that of negroes, a pronounced ellipse.

Example 7.4 The leg bones that break more frequently when compressed are the tibia immediately above the ankle, where the cross-sectional area is about 3 cm^2. Find the compressive load (mass) that causes a fracture of the tibia of one of the legs.

We assume that the tibia has the same elastic properties as the femur.

$T_c = \sigma_c A = (16.7 \times 10^7 \text{ Pa})(3 \times 10^{-4} \text{ m}^2)$

$T_c = 5.0 \times 10^4$ N. This force is equivalent to supporting a load of 5,000 kg = 5 tons.

Exercise 7.5 Find the maximum compressive force that a femur supports before it breaks, in the cases of an adult male and an adult female, knowing that the effective cross-sectional area of the thin part of the femur of a man is 6.5 cm^2 and of a woman, 5.2 cm^2.

7.6 Pressure or Stress on Intervertebral Discs

Intervertebral discs are located between two vertebrae of the spinal column. These discs have a wall called the fibrous annulus that is composed of about a dozen collagen fibers. The fibers of adjacent layers have different orientations that give strength to the discs under shear motion. At the internal part of a disc, there is a nucleus made of a viscoelastic gel composed of about 80 % of water and proteins. A little of this water is lost during the day due to compression of the discs by standing or by walking which is recovered after a night of sleep.

The elasticity of a disc is due to its wall that can degenerate with age or with repeated overloads. The herniation of the wall followed by nucleus extrusion can compress the nerve which is one of the reasons for back pain. The discs, when subject to a very large pressure, can be forced from their normal location, when it is said that a sequestrum occurred.

The region of the column that is subject to the greatest forces in bending over to pick up a heavy object from the ground is the lumbar, as seen through the calculation of Example 6.5. Moreover, the intensity of this force increases very much for incorrect postures.

Measurements of pressure in the intervertebral discs of human beings have been made by A. Nachemson and published as *Disc Pressure Measurements* in Spine 6 93, 1981. For an adult of 70 kg, standing erect, the pressure on a disc between the third and the fourth lumbar vertebra is 5.5 atm $= 5.6 \times 10^5$ Pa. The force applied there results from the weight force of the set trunk/neck/head/upper arms/fore arms/hands equal to 460 N. Using these data we can find the cross-sectional area A of a disc:

$A =$ weight force/pressure $= 460$ N/$(5.6 \times 10^5$ Pa$) = 8.2 \times 10^{-4}$ m$^2 =$ 8.2 cm^2.

From these data, we can calculate the pressure on the same disc assuming that this subject now holds in each hand a 10 kg mass, hence a total mass of 20 kg. In this case the total weight will be $= 660$ N, and we obtain for the pressure, $p = 7.9$ atm. There was a 44 % increase in the pressure.

Nachemson also measured the pressure on the same disc while picking up an object of 200 N weight with both hands and bending correctly the knees. He obtained an average value of approximately 13.0 atm. When the knees are not bent, the pressure reached 35.0 atm during a very small interval of time, which is due to a compressive force of 2,907 N.

In Example 6.5 in which a person lifts a weight of 200 N without bending his knees, we have obtained, using a simple model, for the contact force applied by the sacrum on the last lumbar disc the value of 3,381 N. As the area of this disc is slightly larger than that used above, we can consider it as being 9.0 cm^2. With these data the pressure can be obtained, giving 37.1 atm, which is coherent with the value measured by Nachemson.

The compressive strength, that is, the compressive breaking stress, for any intervertebral disc is, according to Table 7.3, 1.10×10^7 Pa which is equal to 11 N/mm^2 $= 1,100$ N/cm^2. In the case of Example 6.5, in which the applied force on the last lumbar disc is 3,381 N, we obtain for the pressure the value 3,381 N/ 9.0 cm^2 $= 375.7$ N/cm^2. That means that there is a safety factor of 2.9 times to cause a fracture, if the force is uniformly applied on the total area of 9.0 cm^2.

Passing from the cervical part to the thoracic and then to the lumbar, the weight force that the column must support increases, as already discussed in Chap. 1. Since the compressive strength of any intervertebral disc is practically constant with a value of 1,100 N/cm^2, as the weight force increases, the area of the disc must also increase.

This question becomes important, when a person who practices yoga decides to do a posture of sirsha-asana, the headstand posture in which this subject stays upside-down, with the top of his head on the floor. In this posture, almost all of the body weight falls on the cervical vertebrae. For an adult of 70 kg, the whole body mass, less that of the head, weighs 65.2 kg. Hence, the first cervical disc must support a weight of about 652 N. If this posture is achieved with care, it is still difficult to reach the compressive breaking stress of 1,100 N/cm^2 but if, incautiously, this force is applied in an area smaller than 0.59 cm^2, the pressure can reach the threshold of fracture.

But, what can happen to the vertebrae that have compressive strength 5.8 times smaller than that of an intervertebral disc, according to Yamada's measurement?

7.7   Pressure on the Vertebrae

Vertebrae have an external thin coverage of compact bone and the internal part is made of a spongy material called the trabecular. At the ends of the long bones such as the femur and the humerus, trabecular bones are found predominantly. The central long part is made of a compact bone which has a kind of channel in the middle containing the marrow. As trabecular bone is relatively more flexible than compact bone, it can absorb more energy when large magnitude forces act on it during activities such as walking or running or even jumping. However, in terms of strength to fracture, they differ greatly from bone of the compact type.

The limit of fracture of a vertebra measured as compressive strength by Yamada is 87.9 times and 5.8 times smaller than the compressive strength of compact bone (human femur) and intervertebral discs, respectively, as can be seen in Table 7.3. Thus, according to Yamada's data, if the intervertebral discs of the cervical region have enough strength to support the weight of the body during the headstand posture, the cervical vertebrae do not, since their compressive strength is 0.19×10^7 Pa $= 190$ N/cm^2. If the area of the vertebra is about 3 cm^2 and it has to support a weight of 652 N, the stress at the vertebra will be 217 N/cm^2.

We would like to remember again that since Yamada's bone data have been obtained with bones alone, outside the body and without ligaments or muscles, this can be one of the reasons that, in fact, no cervical vertebra breaks when a person executes the headstand posture with care.

7.8   Shear Stress in the Lumbosacral Intervertebral Disc

The spinal column of a normal person standing erect is not straight, when it is observed from the side. Its curvature, defined by seven cervical vertebrae, twelve thoracic and five lumbar, is called cervical lordosis, dorsal kyphosis, and lumbar lordosis, respectively, and is shown in Fig. 7.6.

The curvature of lumbar lordosis is determined by the lumbosacral angle, that is, the angle between the horizontal and the top surface of the sacrum. This angle is 30°, for a normal person standing erect. An anomalous curvature of the lumbar lordosis can be one of the reasons for low back pain and let us see why. Figure 7.7 shows the lumbar region of the spinal column of a person standing erect.

Fig. 7.6 Normal cervical
lordosis, dorsal kyphosis,
and lumbar lordosis of a
person standing erect

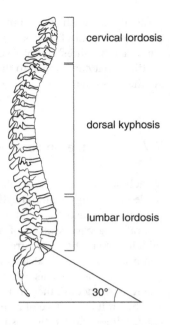

cervical lordosis

dorsal kyphosis

lumbar lordosis

30°

Fig. 7.7 (a) Forces that act
on the lumbosacral disc.
(b) Decomposition of forces
into their orthogonal
components

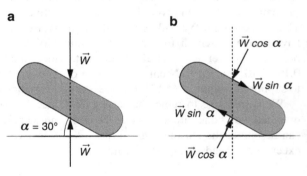

In the erect posture, the weight force that acts on the column is always vertical
and each disc supports everything that is above it and there is also the contact force
of the same value, directed upward, once the body is in equilibrium. Figure 7.7a
shows the weight force W of the set head/neck/trunk/upper arms/fore arms/hands as
the contact force, and the Fig. 7.7b illustrates the decomposition of these forces into
their orthogonal components.

Decomposing the forces, we note that their components act perpendicularly to
the disc, causing its compression, and tangentially to the disc with values $W\sin\alpha$
which causes shear. If the lumbosacral angle of a normal subject standing erect is
30°, the value of the shear force is $W\sin 30° = 0.5\ W$.

If the lumbosacral angle, due to incorrect posture or due to weakness of the
flexor muscles of the pelvis or of the abdominal muscles, increases to 40°, the shear

forces increase to $W\sin 40° = 0.64\ W$, which can be the cause of back pain. It is known that a disc is much less resistant to shear stress than to compressive stress. Hence, among the exercises to alleviate low back pain should be one to decrease the lumbosacral angle.

7.9   Bone Fractures in Collisions

When somebody falls from a springboard into the water, or when he or she collides with a glass door or into something accidentally, what are the types of force subject on them? To respond to this question, let us begin by analyzing the equations that govern the movement of accelerated or decelerated bodies. We will verify that the crucial quantity for this question is the duration of the collision.

An object falling freely, that is, simply dropped from rest at a certain height, falls under the action of gravitational acceleration. The distance H traveled by a body in free fall during an interval of time t can be written as:

$$H = \frac{1}{2}gt^2. \tag{7.6}$$

As already used, g is the acceleration due to gravity. We say that this body undergoes a uniform accelerated motion[3] and the distance traveled increases with the square of the time interval because its velocity increases with time. As the object was released from rest, its initial velocity was equal to zero and the final velocity can be calculated from:

$$v = gt, \tag{7.7}$$

meaning that the velocity increases proportionally with time. Substituting t of (7.6) into (7.7), the final velocity is obtained:

$$v^2 = 2gH; \quad v = \sqrt{2gH}. \tag{7.8}$$

Note that the final velocity does not depend on the mass of the object, i.e., it will be the same for a piece of paper or a heavy lead ball, if the force of the air resistance during its fall is neglected.

---

[3] This is valid if the altitude is not very high, which will be the case here considered. Otherwise, the resistance to the motion by the air increases with the velocity, so that for any falling body there is a limiting velocity for which there is no longer any net acceleration and the velocity ceases to increase, becoming constant.

Similarly, the relation between the average deceleration $\bar{a}$ required to bring a body to rest, from a velocity v, and the distance d to be traveled until it stops is given by:

$$v^2 = 2\bar{a}d. \tag{7.9}$$

As (7.8) and (7.9) are equivalent, we can set them equal and solving for $\bar{a}$ we find:

$$\bar{a} = g\frac{H}{d}. \tag{7.10}$$

The above equations are valid also for a body under constant acceleration, as in the case of a car in motion with constant acceleration instead of the fall from a certain height due to the gravitational acceleration. In this case we can just substitute the acceleration due to the gravity g by $a = \Delta v/\Delta t$, i.e., the change in the velocity in a certain time interval.

The second law of Newton, already presented in Chap. 1, states that:

$$F = ma = m\frac{\Delta v}{\Delta t}. \tag{7.11}$$

After a person jumps from a springboard, for example, during the collision with the water surface, deceleration occurs until its final velocity becomes equal to zero. This deceleration is not constant, but its average value can be obtained, by calculating the difference between the velocity on reaching the water's surface and the final velocity that is zero, i.e., $\Delta\bar{v} = v - 0$, divided by time interval. So, during the collision, the body is subject to a force exerted by the water surface which is also not constant but its average value can be obtained by:

$$\bar{F} = m\frac{\Delta\bar{v}}{\Delta t}, \tag{7.12}$$

or also through:

$$\bar{F} = m\bar{a} = mg\frac{H}{d}. \tag{7.13}$$

The product of the applied force by the time interval during which the force acts is a physical quantity called the impulse I of the force $\bar{F}$:

$$I = \bar{F}\Delta t = m\bar{a}\Delta t = m\Delta\bar{v}. \tag{7.14}$$

Remember that $\Delta\bar{v} = v - 0$, where v is the velocity at the collision and zero, the final velocity, when the body stops.

Example 7.5 A running child hits his head on a glass door because he did not notice that the door was closed. At the instant of the collision, the child's velocity was 3 m/s. The head stops on the glass and its final velocity is zero. Consider that the mass of the child's head is 3.0 kg and the duration of the collision is 0.01 s. Find the average decelerating force exerted by the glass door on the child's forehead.

(a) Using (7.12), we obtain:

$$\bar{F} = (3.0 \text{ kg})(3 \text{ m/s})/(0.01 \text{ s}),$$

$$\bar{F} = 900 \text{ N}.$$

Observe the size of this force, since it is equivalent to laying the child on the floor and placing on his forehead a mass of 90 kg.

Let us now change the glass door to a cushioned door, which dampens the collision. In this case, the time interval of the collision changes to 0.10 s. The force exerted by the cushioned door will be:

$$\bar{F} = (3.0 \text{ kg})(3 \text{ m/s})/(0.10 \text{ s}),$$

$$\bar{F} = 90 \text{ N},$$

one tenth the previous case.

Physical impacts are characterized by a rapid deceleration. The helmets used by motorcyclists are especially designed to dampen the shock to the head at high velocity and increase the duration of the impact to decrease the decelerating forces and to minimize the damage to the head. The role of air-bags is also to increase the collision time that dampens the crash.

Exercise 7.6 In accidents involving a motorcyclist who rides without a helmet, death may result if he receives a blow to his head with an impulse of 100 N s. What should be the maximum velocity of a motorcyclist if he wants to avoid dying? Consider the mass of the motorcyclist's head to be 5.0 kg.

Example 7.6 A woman of 60 kg mass jumps with stiff legs from a table of 1 m height onto a hard floor tile. During the collision, a deceleration to a state of rest occurs in a time interval of 0.005 s. Calculate: (a) the average force exerted on each foot by the ground; (b) the distance traveled by the body during the collision.

We have to begin by calculating the final velocity of fall when the body reaches the ground:

$$v = \sqrt{2gh},$$

$$v = 4.47 \text{ m/s} = 16.1 \text{ km/h}.$$

(a) Using (7.12) and remembering that the variation in velocity is 4.47 m/s, since it decreases from this value to zero when she stops, we obtain:

$$\bar{F} = (60\,\text{kg})(4.47\,\text{m/s})/(0.005\,\text{s}),$$

$$\bar{F} = 53,640\,\text{N},$$

which is about 90 times the weight of her own body. This is the force applied on both feet. Hence, on each foot the applied force would be half:

$$\bar{F} = 26,820\,\text{N}.$$

In Example 7.4 we found that a compressive force of 50,000 N breaks the tibia of a person. Therefore, in this case, the woman is within a safety factor. However, if she were to fall on one leg stiffly, she would not have a safety factor and the fracture would occur. To decrease this force, it is necessary to dampen the fall, for example, by bending the knees during the landing, which increases the deceleration time. That is what athletes do who practice jumping or even football players who know how to fall, rolling to dampen the landing.

(b) $d = gH/(\Delta v/\Delta t) = 0.011$ m; $d = 1.1$ cm

According to Dr. A.S. Iutaka, the bone that fractures more easily, when subject to compression during a fall from a certain elevation with stiff legs, is the calcaneus of the sole of the foot. It is a bone of the trabecular type with a layer of compact bone on its surface.

Exercise 7.7 Find the maximum height from which a person with 100 kg mass and a rigid outstretched leg can fall without breaking the calcaneus. Suppose that in such a condition the damping distance traveled during the collision with a hard ground to a stop is 1.0 cm. Consider the compressive strength of the calcaneus the same as that of a vertebra. Calculate also the duration of the collision. Consider for this person that the area of calcaneus region that touches the ground is 5.5 cm^2.

Exercise 7.8 Find the average deceleration force exerted by the ground on each foot of a person of 70 kg mass during a running activity, in which the height of the fall of one of the feet before it touches the ground is 10 cm. Consider the case of a person who does not know how to dampen the fall and the other who knows. In the first case the interval time of damping is 0.01 s and in the second 0.10 s.

7.10   Answers to Exercises

Exercise 7.1 (a) $T = 10$ kN; (b) $\varepsilon = 0.17$ %; (c) $T = 20$ kN.

Exercise 7.2 (a) $r = 2.82$ cm; (b) $\sigma = 2.8 \times 10^8$ Pa; (c) shortening $= 0.93$ cm; (d) Weight $= 5 \times 10^5$ N.

Exercise 7.3 elongation percentage $= 0.00145$ %.

Exercise 7.4 radius $= 0.44 \times 10^{-4}$ m $= 44$ μm.

Exercise 7.5 $T_M = 1.08 \times 10^5$ N and $T_F = 8.68 \times 10^4$ N.

Exercise 7.6 $v = 72$ km/h.

Exercise 7.7 (a) $H_{max} = 1.0$ cm; (b) $\Delta t = 44$ ms. Here, again, we have obtained a fictional value, due to the value of the compressive strength of the calcaneus (trabecular bone) presented by Yamada. The value does not represent the real situation in which there is interference by muscle and articulations that are not considered in the experimental measurements. If we consider that this bone has the same property as the compact bone, concerning the strength, we obtain a very reasonable value of $H_{max} = 0.92$ m and $\Delta t = 4.7$ ms.

Exercise 7.8 (a) Average force $= 9,900$ N on the sole of the foot that touches the ground; (b) Average force $= 990$ N on the sole of the foot that touches the ground. Observe how important it is to dampen the fall.

Chapter 8
Experimental Activities

The physical concepts presented in this book can be explored through experimental activities, performed with specific materials in a teaching laboratory or with alternative materials in the classroom. Here we have made a choice for alternative materials.

8.1 Objectives

- To perform at least one experiment related to each topic presented in Chaps. 1–7, for a better understanding and memory retention of the concepts.

8.2 Introduction

Before the presentation of the proposals for experimental activities related to the chapters of this book, we have to learn how to write correctly the results of a measurement.

8.2.1 Significant Digits

The first difficulty in an experimental activity is the measurement itself. When a measurement is performed, employing the same methods and instruments in conditions considered to be the same, in general, discordant results are obtained. Such fact is justified, by noting that measurements are affected by observational uncertainties due to factors such as the quality of the apparatus and the skill of the experimenter. In these circumstances, one need to answer which number should be adopted as a measure of the quantity and which value will best represent it.

E. Okuno and L. Fratin, *Biomechanics of the Human Body*, Undergraduate Lecture Notes in Physics, DOI 10.1007/978-1-4614-8576-6_8,
© Springer Science+Business Media New York 2014

Fig. 8.1 Measurement of
the length of an object with
a ruler with the smallest
division in centimeters

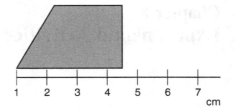

To begin this discussion, let us ask: how to determine the length of the object of Fig. 8.1 with a meter stick marked in centimeters? How many digits should be used to write the result? Which of them are the significant digits?

Measure and write down the length of the object of Fig. 8.1.

Possible readouts:

4.4 cm
4.5 cm
4.6 cm

Incorrect readouts:

4 cm
4.55 cm

In these measurements the first reliable digit is the number 4. Now, the tenth of a centimeter is said to be a doubtful digit, since there are no divisions in the scale to serve as reference, and depends on the evaluation of the experimenter. Hence, the first three readouts are correct if, to perform the measurement, one has in mind that the last digit is an uncertain, unreliable digit. The fourth readout, done without any decimal digit and the fifth, with two decimal digits, are both wrong, in the first case due to the lack and in the second case due to the excess of digits. This allows the establishment of the correct form to express the result of a measurement. Therefore, the result of a measurement must be written with all significant digits, composed of reliable digits and only one estimated digit.

Now, do the same measurement with a meter stick, marked in millimeters, as shown in Fig. 8.2:

Possible readouts:

4.51 cm
4.53 cm
4.58 cm

Incorrect readouts:

4.5 cm
4.572 cm

Now the reliable digits are 4 and 5 and the hundredth of a centimeter is a doubtful digit, which should be written, because it is also significant. The fourth

Fig. 8.2 Measurement of
the length of an object with
a ruler with the smallest
division in millimeters

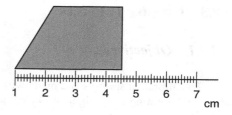

and the fifth readouts are incorrect due to the lack of a digit in the first case and to the excess of digits, in the second case. If the length of the object is between 4.0 and 4.1 cm, the measurement should be written 4.00 or 4.04 or 4.08 cm, for example, with all significant digits. Observe the importance in using apparatus with great precision, but it is still more important to write correctly the performed readouts. Very expensive, precise equipment is valueless if the experimenter does not know such rules.

We should remember that the digit zero is only significant if it is located to the right of a significant digit. Then,

0.00030 has only two significant digits, which are 3 and 0
350,000 has six significant digits

8.2.2 Mathematical Operations with Significant Digits: Two Examples

Sum
34.4
0.006
115
149.406

The correct form to write the result of sum is 149, since the unit is a doubtful digit because it has been estimated. It is purposeless to write decimal digits in the result, due to the number 115. The result was rounded off to three significant digits.

Multiplication
3.14 × 1.0
3.1

In this case, the result of multiplication cannot have more significant digits than the quantity of digits in the least significant digit used in the calculation (1.0).

8.3   Chapter 1: Forces

8.3.1   Objectives

- To calibrate a spring scale made with a rubber band
- To determine the weight of objects and the magnitude of forces present in an experimental system
- To establish the conditions of static equilibrium relative to translation

8.3.1.1   Necessary Materials

- Rubber band
- Paper clips
- Objects with known mass
- 30 cm ruler or a tape measure
- Protractor

The objects of known mass can be a 1.0 or 2.0 l plastic bottle with a known quantity of water. Remember that the mass of water corresponding to a volume of 1.0 l is 1.0 kg[1] and, hence, represents a weight of 10 N. If it is not possible to determine the volume of water in known units, work with an arbitrary unit of mass, a.u.m., adopting a certain volume of water such as a glass of water or a soda can as a standard.

8.3.2   Activity 1: Construction and Calibration of a Spring Scale (Dynamometer)

The word dynamometer originates from the Greek dynamis. It is an apparatus developed to measure the magnitude of a force. A rubber band or a spring can be used to construct a spring scale. In the case of a spring, we can apply Hooke's law that states: the change in the length, Δx, observed in a spring is directly proportional to the magnitude of the force, F, applied to it, that is, if we double the magnitude of the force, the change in the length of the spring is doubled and so on. This proportionality can be expressed by (8.1):

$$F = k\Delta x, \tag{8.1}$$

---

[1] 1.0 l of water $= 1,000$ cm^3. Considering the density of water $= 1.0$ g/cm^3, we obtain for the mass of water $= 1,000$ g $= 1.0$ kg.

where F is the magnitude of the force (in newtons) exerted by the spring, Δx is the observed variation in the spring length (in meters), and k is the elastic constant of the spring, measured in N/m.

In the case of a rubber band, used here, the change in the length, Δx, of the elastic cannot be directly proportional to the magnitude of the applied force, F, because the profile of the elastic, that is, its cross-sectional area changes significantly with the increase of the force. Therefore, the recommended procedure is to construct a calibration curve that relates F to Δx. Whenever a rubber band is used as a spring scale, one has to measure initially the increase in the length of the elastic for a known force and from the graph of F versus Δx determine the corresponding magnitude of the applied force.

8.3.2.1 Procedure

Construction of a spring scale (dynamometer)

1. Cut a rubber band to obtain a strip.
2. Redo two paper clips, as shown in Fig. 8.3, to transform them into hooks.
3. Tie one clip at each extremity of the rubber band and the spring scale is ready.

Calibration of the Spring Scale

4. Measure the length, x_o, of the outstretched elastic of the rubber band spring scale, from end to end, in meters.
5. Make a small hole near the mouth of a plastic bottle, as shown in Fig. 8.4, to hook one of the clips of the spring scale.
6. Pour one unit of mass into the bottle and measure the length of the stretched rubber band strip, x_1, in meters.
7. Determine the increase in the length of the rubber band strip, by calculating the difference $x_1 - x_o$.
8. Fill out Table 8.1 with the value of the force, F, exerted on the rubber band (in newton, N, or in an arbitrary unit of force, a.u.f.) and the corresponding Δx (in meter, m) obtained through calculation.
   Consider: 1 N as the weight force that acts on a mass of 100 g,
   or: 1 a.u.f. is the weight force that acts on 1 a.u.m.
9. Repeat procedures 6, 7, and 8 submitting, successively, the spring scale to two, three, four and five units of mass.
10. Construct the calibration curve of the spring scale, that is, the graph of the magnitude of the applied force, F, as a function of the corresponding observed change in the length, Δx, of the rubber band.
11. Determine the weight and the mass of an object by hanging it in place of the pet bottle and measuring Δx. Then, use the spring scale calibration curve.
12. $W = $ _____(N or a.u.f.) $m = $ _____ (kg or a.u.m.).
13. Construct and calibrate another spring scale made with a rubber band strip.

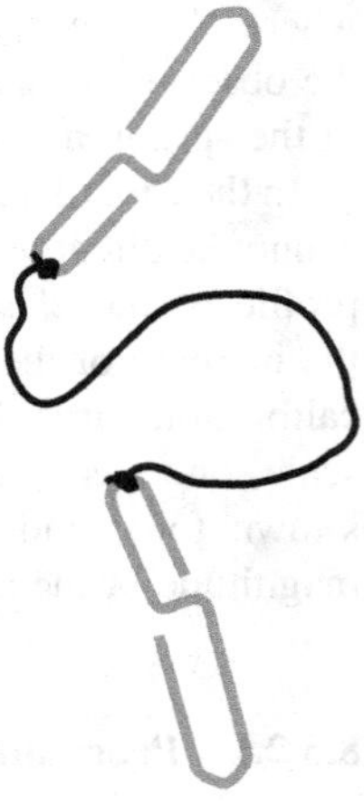

Fig. 8.3 Spring scale
constructed with a rubber
band and paper clips
attached to its ends

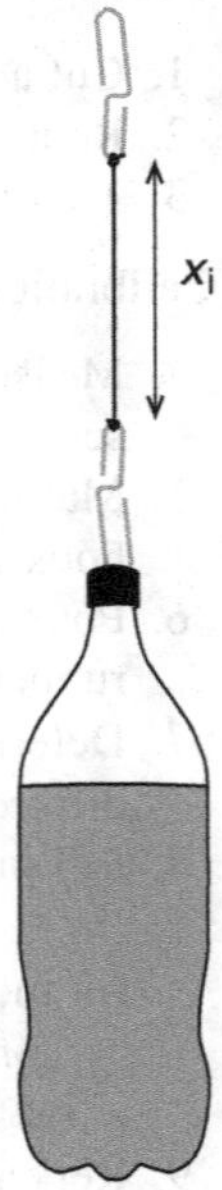

Fig. 8.4 A pet bottle with
water supported by a rubber
band spring scale

Table 8.1 Applied force
on the spring scale made of
a rubber band and the
corresponding change in
its length

F (N or a.u.f.)	Δx (m)

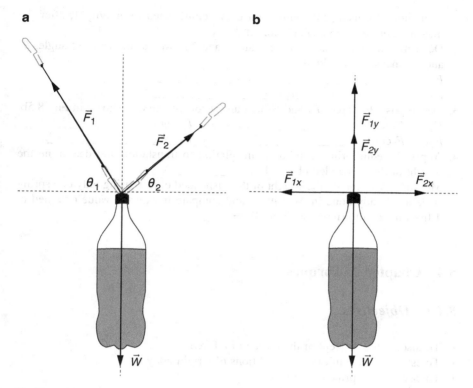

Fig. 8.5 (**a**) Object of unknown mass in static equilibrium, supported by two rubber band strip spring scales whose angles with the horizontal are θ_1 and θ_2. (**b**) Decomposition of involved forces in static equilibrium

8.3.3   Activity 2: Conditions for the Static Equilibrium with Respect to Translation

The static equilibrium of a suspended object subject to three coplanar forces shown in Fig. 8.5 will be established. The magnitude of these forces will be determined, by using the spring scales previously constructed and calibrated in Activity 1. We shall verify, using decomposition of forces, if the static equilibrium conditions in relation to translation are satisfied.

8.3.3.1   Procedure

1. Suspend an object with known mass by using two spring scales as illustrated in Fig. 8.5a. In this figure, F_1 is the result of the readout of spring scale 1 and F_2, of spring scale 2. The result of the readout means the value of the magnitude

F obtained by using the calibration curve constructed in activity 1, after the measurement of x_i and calculation of Δx.

2. Determine the readouts of spring scales 1 and 2 as well as the value of angles θ_1 and θ_2, measured by the protractor.

$F_1 = $ _____ $\theta_1 = $ _____

$F_2 = $ _____ $\theta_2 = $ _____

3. Decompose the forces F_1 and F_2 in x and y components, as shown in Fig. 8.5b.

$F_{1X} = F_1\cos\theta_1 = $ _____ $F_{1Y} = F_1\sin\theta_1 = $ _____

$F_{2X} = F_2\cos\theta_2 = $ _____ $F_{2Y} = F_2\sin\theta_2 = $ _____

4. Apply the equilibrium conditions in relation to translation and determine the weight of the suspended object. $W = $ _____

5. Measure now directly the weight of the suspended object, using only one spring scale as already done in Activity 1, and compare it with the value obtained in 4 from the equilibrium condition. $W = $ _____

8.4   Chapter 2: Torques

8.4.1   Objectives

- To discuss the concept of the torque of a force
- To establish the equilibrium conditions of a rigid body
- To develop the principle of lever

8.4.2   Activity 3: Torque of a Force

8.4.2.1   Necessary Materials

- A wooden bar or equivalent
- Triangular support or equivalent
- Objects with known mass (30 small marbles with diameter of around 1.2 cm, for example)
- Devices to hold the masses (plastic cup or equivalent)
- 30 cm ruler or a measuring tape

8.4.2.2   Procedure

1. Establish static equilibrium and measure the masses and distances of the experimental arrangement of Fig. 8.6, in the following situations:

Fig. 8.6 Experimental arrangement used to establish static equilibrium in the conditions proposed in this activity. A plank is maintained in equilibrium, placing it on the triangular support base

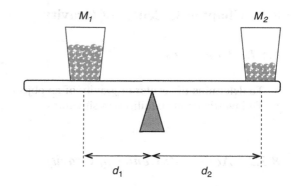

$M_1 = M_2 = 4$ marbles, for example;
$M_1 = 2 M_2$;
$M_1 = 3 M_2$;
and fill out Table 8.2.

Table 8.2 Data obtained with the experimental setup used to establish static equilibrium in the conditions proposed in this activity

Situation 1

F_1 (N)	d_1 (m)	F_2 (N)	d_2 (m)	$F_1 d_1$ (N m)	$F_2 d_2$ (N m)

Situation 2

F_1 (N)	d_1 (m)	F_2 (N)	d_2 (m)	$F_1 d_1$ (N m)	$F_2 d_2$ (N m)

Situation 3

F_1 (N)	d_1 (m)	F_2 (N)	d_2 (m)	$F_1 d_1$ (N m)	$F_2 d_2$ (N m)

If necessary, use a.u.f. in place of newton, N, for the unit of force

8.4.2.3 Static Equilibrium

1. The necessary conditions for static equilibrium of a rigid body were already discussed in Chap. 6, that is, in summary: $\vec{R} = \sum \vec{F} = 0$, in translation and, $M_T = \sum M_F = 0$, in rotation.
2. Verify, mathematically, if the conditions for equilibrium relative to rotation are observed in each situation. For evaluation, consider the correct number of significant digits in each measurement performed.
3. Determine the weight of the wooden bar, using the rubber band spring scale constructed in Activity 1.
4. Apply the conditions for equilibrium relative to translation and determine the magnitude of the support force (exerted by the triangular base on the wooden bar) in each situation.
5. Justify the fact that the weight of wooden bar was not taken into account for the verification of the rotational equilibrium conditions.

8.5 Chapter 3: Center of Gravity

8.5.1 Objectives

- To determine the center of gravity of an object
- To investigate the equilibrium situations

8.5.2 Activity 4: Center of Gravity

8.5.2.1 Necessary Materials

- Two large paper clips
- Needle-nose pliers
- Tape
- Two coins

8.5.2.2 Procedure

Construct a structure equivalent to that of Fig. 8.7, with the suggested material. Open one of the paper clips using a needle-nose pliers and transform it into a new V form. Join the two clips with tape, as illustrated in Fig. 8.7. Then, fix each coin with tape at the extremity of the clip in V (upside-down). The structure is ready. Now rest it on a pencil or a pole and verify the equilibrium. Think about the concept of center of gravity and of torque due to the weight force on a body and justify the observed equilibrium.

You can use other material and elaborate other forms to investigate the center of gravity of these forms and the importance of torque of weight force.

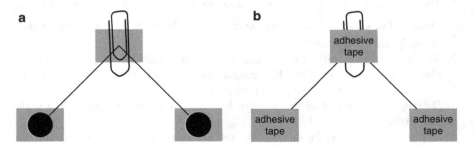

Fig. 8.7 Structure elaborated for the investigation of the concept of center of gravity of an object and the corresponding torque due to the weight force. (**a**) Front view and (**b**) back view of the structure

8.6   Chapter 4: Rotations

8.6.1   Objectives

- To investigate the physical quantities moment of inertia and angular momentum
- To analyze the conservation of angular momentum in rotational motion
- To relate the change in the angular velocity to modifications in the mass distribution of a rotational system

8.6.2   Activity 5: Moment of Inertia and Angular Momentum

8.6.2.1   Necessary Materials

- 4 large paper clips that will play the role of distributed masses
- Rubber band strip
- Tape
- Sewing thread
- Scissors
- Pliers
- Rigid metallic wire

8.6.2.2   Procedure

1. Cut two pieces of rigid metallic wire with an equal length of approximately 20 cm. The length of the rubber band strip must be longer than the length of the clip.
2. Construct the arrangement of Fig. 8.8 with the metallic wires.

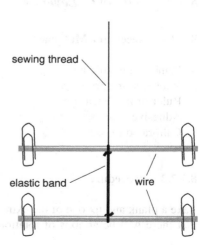

Fig. 8.8 Structure for investigation of the concept of moment of inertia and angular momentum

3. Tie the sewing thread to one of the metallic wires so that the arrangement can be suspended. The strip of rubber must be tied to each metallic wire in the middle as shown in Fig. 8.8. The sewing thread and the rubber strip must be tied in a way that the structure stays in equilibrium when released, and can rotate freely.
4. Fix one large clip near the extremity of each metallic wire. Ensure that the mass of both metallic wires with clips is the same.
5. Rotate one of the metallic wires clockwise and the other in the counterclockwise direction, twisting the rubber.
6. Release the arrangement and report what you observe.
7. Change the position of both clips closer to the rubber, in one of the metallic wires, keeping the other wire as it is. Observe that the masses of each wire with two clips remain the same. Repeat the procedures 5 and 6.
8. Do you observe a change in the rotational velocity? Why?
9. Explain the reason for the opposite direction of the rotations of wires. Observe that the arrangement is at rest when it is freed.

Note: You can improve the experiment, increasing the number of clips.

8.7 Chapter 5: Simple Machines

8.7.1 Objectives

- To apply the principle of the lever and the equilibrium conditions of a rigid body
- To classify levers
- To analyze devices that allow the discussion of mechanical advantage

8.7.2 Activity 6: Levers

8.7.2.1 Necessary Materials

- Plank or metallic bar
- Rubber strip spring scale
- Ruler or tape measure
- Adhesive tape
- Calibrated masses

8.7.2.2 Procedure

Take a plank and fix one of the extremities to the workbench with adhesive tape so that there will be an axis of rotation. Use the adhesive tape to form two support

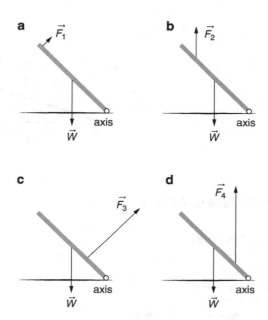

Fig. 8.9 Representation of the proposed activity in which the wooden plank is maintained in static equilibrium through the torque of applied forces F and measured by a spring scale, in four different situations: (**a**), (**b**), (**c**) and (**d**). The weight force W is represented at the plank's center of gravity. The point of application of forces F corresponds to the loops made with tape attached to the plank. The rotation axes correspond to the point where the plank is fixed with the tape to the workbench

loops, placing one near the free extremity as shown in Fig. 8.9a, b and another, at any place between the middle of the plank and the axis of rotation as illustrated in Fig. 8.9c, d. If the mass of the plank is very small, add a mass by attaching it with adhesive tape at its center of gravity.

1. To establish static equilibrium in the situations shown in Fig. 8.9, apply the necessary force. For this, use the rubber band spring scale and its calibration curve.
2. For each of the situations:

   (a) Measure the force on the spring scale and determine the arms of action and of resistance (plank weight) force. Consider that the plank is homogenous and its center of gravity is in its center.
   (b) Determine from the static equilibrium conditions:

   – The value of the weight of the plank
   – The magnitude of the support force (normal reaction) for situations b) and d)
   – The type of lever used

Fig. 8.10 Illustration
of an inclined plane

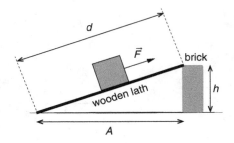

8.7.3   Activity 7: Inclined Plane

8.7.3.1   Necessary Materials

– Rubber band spring scale
– Block of wood or equivalent
– Tape measure or ruler
– Wooden lath with around 50 cm length
– Brick or equivalent
– Adhesive tape

8.7.3.2   Procedure

1. Determine the weight of the block. For this, use the rubber band spring scale.
2. Use the wooden lath and the brick to construct the inclined plane, as shown in Fig. 8.10. The slope of inclined plane is fixed by the height of the brick.
3. Measure A, d, and h.
4. Attach the rubber band spring scale to the wood block and pull it up the ramp, moving it continuously, in a uniform way, in order to maintain a constant velocity. Determine the value of the force exerted for this.
5. Find the ratio between the value of the weight of the block and the value of the force to pull it up the ramp and determine the mechanical advantage.
6. Determine the theoretical mechanical advantage by calculating the ratio between the lengths d and h. Try to justify the possible difference.
7. Change the height h, by using a different dimension of the brick, and repeat the items from 3 to 6.
8. Compare the results obtained in 6 and 7 and discuss.

8.8 Chapter 6: Muscle Force

8.8.1 Objectives

- To characterize the association of rubber bands
- To establish the analogy between the association of rubber bands and the configuration of muscle fibers

8.8.2 Activity 8: Association of Springs

8.8.2.1 Necessary Materials

- Rubber band strip
- Paper clips
- Ruler or tape measure
- Known masses
- Support of masses

8.8.2.2 Association of Springs: In Series and in Parallel

We have already mentioned that in a spring, the change observed in its length is proportional to the force applied to it (Hooke's law). The constant k of proportionality is called the force constant of the spring. There are two ways to associate springs: in series and in parallel.

In the association of N springs *in series*, one following the other in a linear sequence, the force on each spring is the same and we observe the following relation for the resultant elastic force constant k_S:

$$\frac{1}{k_S} = \frac{1}{k_1} + \frac{1}{k_2} + \cdots + \frac{1}{k_N}.$$

Observe that if the springs are identical, we can write: $k_1 = k_2 = \cdots = k_N = k$ and, hence, $k_S = \frac{k}{N}$.

In the association of N springs *in parallel*, one beside the other, the change in the length of each spring is the same and we observe the following relation for the resultant force constant k_P:

$$k_P = k_1 + k_2 + \cdots + k_N.$$

Observe now that if the springs are identical, $k_P = Nk$

Fig. 8.11 Association in
series of three rubber strip
spring scales

8.8.2.3 Association of Springs

In this experiment the rubber strip is used. Its behavior is not the same as a spring, as
the relation between the magnitude of the applied force, F, and Δx is not constant,
but, in any case, there is a relation between these quantities.

As in the case of springs, we can associate rubber strips in series and in parallel
and quantify the relation between Δx and applied F for the association.

8.8.2.4 Procedure

1. Construct three spring scales identical to those of Activity 1 of this chapter.
   Assume that the rubber strip of each apparatus, being of the same length,
   obeys the same calibration curve relating F and Δx, already determined for
   that spring scale.

Association in Series

2. Use three spring scales in a series association as illustrated in Fig. 8.11.
3. Measure the initial length, x_0, of association, without any applied force. You can
   measure the length of associations, considering the distance between the extremity of the first clip and the extremity of the last clip. Thus, the measured length is
   6 clips + 3 rubber strips.
4. Submit the association in series of the spring scales to a known force (it can be a
   pet bottle with water) and determine the increase in the length of the association.
   Write the result in Table 8.3. As the length of the clip does not change, what is
   measured now is the length of 6 clips plus 3 elongated rubbers. When Δx is

Table 8.3 Force applied
to the association of rubber
strip spring scales in series
and the corresponding
variation in the length of the
association

F (N or a.u.f.)	Δx (m)

Fig. 8.12 Association of
three rubber strip spring
scales in parallel. Observe
that the three hooks should
be assembled to constitute
together a single hook

Table 8.4 Applied force to
the rubber strip spring scale
association in parallel and the
corresponding elongation of
the association

F (N or a.u.f.)	Δx (m)

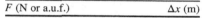

calculated, the lengths of clips are canceled, with only the change in the length of
the rubber strips remaining.

5. Repeat the previous procedure for other values of mass and construct the
   calibration curve for the association.
6. Compare the calibration curve of a single spring scale with that of the association in series. What do you conclude?

Association in Parallel

7. Associate three spring scales in parallel as illustrated in Fig. 8.12.
8. Measure the initial length, x_o, of the association without an applied force.
9. Submit the association to a known force and determine the elongation of the
   association. Write the result in Table 8.4.
10. Repeat the previous procedure for other values of mass and construct the
    calibration curve for this association.
11. Compare the calibration curve of a single spring scale and that of the associa-
    tion in parallel. What do you conclude?
12. Try to establish an analogy between this type of association and how muscle
    fibers are grouped.

8.9   Chapter 7: Bones

8.9.1   Objectives

- To illustrate the relation between the stress, σ, applied to a material and the strain, ε, observed in it
- To illustrate the tensile strength of material

8.9.2   Activity 9: Strength of a Spring

8.9.2.1   Necessary Materials

- Rubber strip spring scale
- Ruler
- Pet bottle with 2 l volume
- Water

8.9.2.2   Procedure

1. The cross-sectional area of the rubber that constitutes the spring scale decreases as the rubber is submitted to increasing force. However, we will consider it constant. Measure the cross-sectional area in m^2, before the application of any force.
2. Make a small hole near the mouth of the pet bottle in order to hook the clip of the rubber strip spring scale, as shown in Fig. 8.4 of Activity 1.
3. Submit the rubber strip spring scale to increasing weight by adding water to the bottle until the rubber breaks. Write the values of the weights and the respective increase in the rubber length, until its rupture. For this, you should control the volume of added water. Plot the graph of stress, σ, as a function of strain, ε, and compare it with the graph of Fig. 7.5. Remember that stress $\sigma = T_{applied}/A =$ weight/area and the strain $\varepsilon = \Delta x/x_o$.
4. Determine the tensile strength at the point of rupture of the rubber strip.

Appendix

Some Mathematical Relations Used in This Book

Geometry

A few examples of supplementary angles are shown in Fig. A.1 for review:

Theorem *If the corresponding sides of two angles are perpendicular to each other, the angles are equal. In Fig. A.2, angle α is formed by sides* a *and* b. *As side* c *is perpendicular to side* b *and* d, *perpendicular to side* a, *the angle formed by sides* c *and* d *is also* α.

Theorem *The sum of the interior angles of a triangle equals 180°, as shown in Fig. A.3.*

Trigonometry

In a right triangle ABC (Fig. A.4), the following trigonometric functions are defined:

$$\sin \alpha = \frac{\text{opposite side}}{\text{hypotenuse}} = \frac{a}{c} \quad \text{and} \quad \alpha = \arcsin \frac{a}{c},$$

$$\cos \alpha = \frac{\text{adjacent side}}{\text{hypotenuse}} = \frac{b}{c} \quad \text{and} \quad \alpha = \arccos \frac{b}{c},$$

$$\tan \alpha = \frac{\sin \alpha}{\cos \alpha} = \frac{\text{opposite side}}{\text{adjacent side}} = \frac{a}{b},$$

$$\sin^2 \alpha + \cos^2 \alpha = 1.$$

E. Okuno and L. Fratin, *Biomechanics of the Human Body*, Undergraduate Lecture Notes in Physics, DOI 10.1007/978-1-4614-8576-6, © Springer Science+Business Media New York 2014

Fig. A.1 Supplementary
angles

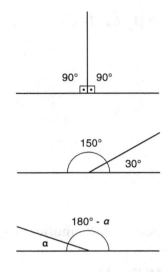

Fig. A.2 Two angles with
corresponding sides
perpendicular to each other

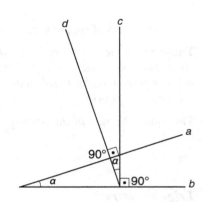

Fig. A.3 The interior
angles of a triangle

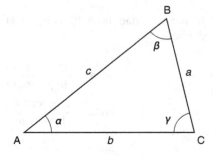

Fig. A.4 A right triangle

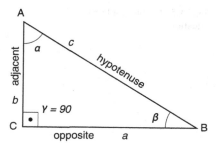

Table A.1 Trigonometric functions of some angles

Angle (deg)	Angle (rad)	sine	cosine	tangent
0	0	0	1.000	0
30	$(1/6)\pi$	0.500	0.866	0.577
45	$(1/4)\pi$	0.707	0.707	1.000
60	$(1/3)\pi$	0.866	0.500	1.732
90	$(1/2)\pi$	1.000	0	∞
180	π	0	−1.000	0
270	$(3/2)\pi$	−1.000	0	$-\infty$
360	2π	0	1.000	0

In a right triangle (Fig. A.4), the Pythagorean theorem can be expressed as in Pythagorean equation:

$$c^2 = a^2 + b^2.$$

In any triangle ABC, shown in Fig. A.3, the following relations are defined:

$$\sin\frac{\alpha}{a} = \sin\frac{\beta}{b} = \frac{\sin\gamma}{c},$$

$$c^2 = a^2 + b^2 - 2ab\cos\gamma.$$

If γ is equal to 90°, which corresponds to the case of a right triangle, the above equation degenerates into the Pythagorean formula, as $\cos 90° = 0$. The values of sine, cosine, and tangent of some angles are given in Table A.1.

Linear Equation

A linear equation has the general form

$$y = a + bx,$$

where a and b are constants. This equation is referred to as being linear because the graph of y vs. x is a straight line as shown in Fig. A.5. The constant a, called the

Fig. A.5 Graph of a linear
equation

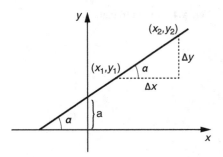

intercept, is the value of y at which the straight line intersects the y-axis. The constant b represents the slope of the straight line and is the tangent of the angle that the line makes with the x-axis. The slope of the straight line is expressed by

$$\text{slope} = \frac{y_2 - y_1}{x_2 - x_1} = \frac{\Delta y}{\Delta x} = \tan \alpha.$$

Observe that a and b can have either positive or negative values. If b is positive (negative), the straight line has a positive (negative) slope. When the straight line passes through the origin, $a = 0$; straight lines parallel to the x-axis have the same slope $b = 0$ and different values of a.

References

1. G.B. Benedek, F.M.H. Villars, *Physics with Illustrative Examples from Medicine and Biology, Three volumes*, 2nd edn. (AIP, New York, NY, 2000)
2. O.H. Blackwood, W.B. Herron, W.C. Kelly, *Física na escola secundária* (INEP/MEC, Rio de Janeiro, 1958)
3. C.S. Calçada, J.L. Sampaio, *Física Clássica: dinâmica e estática*, vol. 2 (Editora Atual, S. Paulo, 2000)
4. J.R. Cameron, J.G. Skofronick, *Medical Physics* (Wiley, New York, NY, 1978)
5. J.R. Cameron, J.G. Skofronick, R.M. Grant, *Physics of the Body*, 2nd edn. (Medical Physics Publishing, Madison, WI, 1999)
6. A.H. Cromer, *Physics for the Life Sciences*, 2nd edn. (McGraw-Hill Co., New York, NY, 1977)
7. H.Q. Fuller, R.M. Fuller, R.G. Fuller, *Physics including Human Applications* (Harper and Row, New York, NY, 1978)
8. Y.C. Fung, *Biomechanics – Mechanical Properties of Living Tissues* (Springer, Berlin, 1981)
9. Grupo de Reelaboração do Ensino de Física, *Física 1: mecânica/GREF* (Editora da Universidade de São Paulo, 1990)
10. A.C. Guyton, J.E. Hall, *Human Physiology and Mechanisms of Disease*, 6th edn. (Saunders, Philadelphia, PA, 1997)
11. G.J. Hademenos, *Schaum's Outline of Physics for Pre-med, Biology, and Allied Health Students* (McGraw-Hill, New York, NY, 1998)
12. S.J. Hall, *Biomecânica Básica* (Guanabara-Koogan, Rio de Janeiro, 1991)
13. D. Halliday, R. Resnick, *Fundamentos de Física 1. Mecânica* (Livros Técnicos e Científicos, Rio de Janeiro, 1991)
14. I.P. Herman, *Physics of the Human Body* (Springer, New York, NY, 2007)
15. R.K. Hobbie, J.R. Bradley, *Intermediate Physics for Medicine and Biology*, 4th edn. (Springer, New York, NY, 2007)
16. F.J. Kottke, J.F. Lehmann, Tratado de Medicina Física e Reabilitação, vol. 2, 4a edição (Editora Manole Ltda, S. Paulo, 1994)
17. G.L. Lucas, F.W. Cooke, E.A. Friis, *A Primer of Biomechanics* (Springer, New York, NY, 1999)
18. J.B. Marion, *General Physics with Bioscience Essays* (Wiley, New York, NY, 1979)
19. H.J. Metcalf, *Topics in Classical Biomechanics* (Prentice Hall, Englewood Cliffs, NJ, 1980)
20. A. Nachemson, Towards a better understanding of back pain: a review of the mechanics of the lumbar disk. Rheumatol Rehabil **14**, 129 (1975)
21. B.M. Nigg, B.R. MacIntosh, J. Mester (eds.), *Basic Biomechanics of the Musculoskeletal System* (Human Kinetics, Champaign, IL, 2000)

E. Okuno and L. Fratin, *Biomechanics of the Human Body*, Undergraduate
Lecture Notes in Physics, DOI 10.1007/978-1-4614-8576-6,
© Springer Science+Business Media New York 2014

22. E. Okuno, C. Chow, I.L. Caldas, *Física para Ciências Biológicas e Biomédicas* (Harbra, S. Paulo, 1982)
23. N. Özkaya, M. Nordin, *Fundamentals of Biomechanics – Equilibrium, Motion and Deformation*, 2nd edn. (Springer, New York, NY, 1999)
24. R.A. Serway, *Physics for Scientists and Engineers with Modern Physics*, 3rd edn. (Saunders Golden Sunburst Series, Philadelphia, PA, 1992)
25. L.K. Smith, E.L. Weiss, L. Lehmkuhl, *Cinesiologia Clínica de Brunnstrom* (Editora Manole, S. Paulo, 1997)
26. G. Wolf-Heidegger, *Atlas de Anatomia Humana*, 4th edn. (Guanabara-Koogan, Rio de Janeiro, 1981)
27. H. Yamada, *Strength of Biological Materials* (Williams and Wilkins, Baltimore, MD, 1970)
28. V.M. Zatsiorsky, W.J. Kraemer, *Science and Practice of Strength Training*, 2nd edn. (Human Kinetics, Champaign, IL, 2006)

Abbreviations

Greek Symbols Used in This Book

Lower case	Capital letter
α	
β	
γ	
δ	Δ
ε	
θ	Θ
μ	
π	
ρ	
σ	Σ
ϕ	Φ
ω	Ω

Other Symbols Used in This Book

Symbol	Quantity	Symbol of unit
F	Force (magnitude or intensity)	N (newton)
a	Acceleration	$m/s^2 = m\ s^{-2}$
W	Weight	N (newton)
p	Pressure	Pa (pascal)
M	Torque	N m

(continued)

E. Okuno and L. Fratin, *Biomechanics of the Human Body*, Undergraduate
Lecture Notes in Physics, DOI 10.1007/978-1-4614-8576-6,
© Springer Science+Business Media New York 2014

(continued)

Symbol	Quantity	Symbol of unit
m	Mass	kg
I	Moment of inertia	kg m^2
α	Angular acceleration	$1/s^2 = 1\ s^{-2}$
f	Rotational frequency	Hz (hertz), $1\ Hz = 1\ s^{-1}$
ω	Angular velocity	rad/s $= s^{-1}$
v	Linear velocity	m/s $=$ m s^{-1}
Δt	Time interval	s (second)
K	Kinetic energy	J (joule)
K_{ROT}	Rotational kinetic energy	J (joule)
g	Acceleration of gravity	m/s^2 $=$ m s^{-2}
f_s	Maximum force of static friction	N (newton)
f_k	Force of kinetic friction	N (newton)
μ_s	Coefficient of static friction	Dimensionless
μ_k	Coefficient of kinetic friction	Dimensionless
ρ	Density	kg/m^3 $=$ kg m^{-3}
L	Angular momentum	kg m^2/s $=$ kg m^2 s^{-1}
τ	Work	J (joule)
MA	Mechanical advantage	Dimensionless
σ	Stress	N/m^2 $=$ N m^{-2} $=$ Pa
A	Area	m^2
σ_t	Tensile stress	N/m^2 $=$ N m^{-2} $=$ Pa
σ_c	Compressive stress	N/m^2 $=$ N m^{-2} $=$ Pa
ε	Strain	Dimensionless
Y	Young's modulus	Pa (pascal)
S	Shear modulus	Pa (pascal)
I	Impulse	N s

Index

E. Okuno and L. Fratin, *Biomechanics of the Human Body*, Undergraduate
Lecture Notes in Physics, DOI 10.1007/978-1-4614-8576-6,
© Springer Science+Business Media New York 2014